History of Physics

The Springer book series *History of Physics* publishes scholarly yet widely accessible books on all aspects of the history of physics. These cover the history and evolution of ideas and techniques, pioneers and their contributions, institutional history, as well as the interactions between physics research and society. Also included in the scope of the series are key historical works that are published or translated for the first time, or republished with annotation and analysis.

As a whole, the series helps to demonstrate the key role of physics in shaping the modern world, as well as revealing the often meandering path that led to our current understanding of physics and the cosmos. It upholds the notion expressed by Gerald Holton that "science should treasure its history, that historical scholarship should treasure science, and that the full understanding of each is deficient without the other." The series welcomes equally works by historians of science and contributions from practicing physicists.

These books are aimed primarily at researchers and students in the sciences, history of science, and science studies; but they also provide stimulating reading for philosophers, sociologists and a broader public eager to discover how physics research – and the laws of physics themselves – came to be what they are today.

All publications in the series are peer reviewed. Titles are published as both print- and eBooks. Proposals for publication should be submitted to Dr. Angela Lahee (angela.lahee@springer.com) or one of the series editors.

Sandra Linguerri · Raffaella Simili

Albert Einstein—Italian Memories

The Bologna Lectures and Other Events

Sandra Linguerri
Department of Philosophy
University of Bologna
Bologna, Italy

Raffaella Simili
Department of Philosophy
University of Bologna
Bologna, Italy

ISSN 2730-7549   ISSN 2730-7557 (electronic)
History of Physics
ISBN 978-3-031-52952-8   ISBN 978-3-031-52950-4 (eBook)
https://doi.org/10.1007/978-3-031-52950-4

Foreword

This is the first book on Albert Einstein which publication the Italian Physical Society (SIF) contributes to. The decision of SIF to contribute to the publication of this book on the relationship of Albert Einstein with Italy is based on the uniqueness of the subject and on the high competence of the authors.

Indeed, in spite of the existing rich literature on this very famous scientist, up to now very little was known on his connections and visits to Italy and on the various memories associated to them.

The first edition of this book, written in Italian and which dates back to 2008, has attracted interest and attention up to recent time as testified by the realization of the documentary "Einstein in Italia" produced and broadcast by RAI Storia in 2021. SIF has contributed also to this initiative and after the documentary SIF has decided to share this important scientific and historical information with the international audience. This motivated SIF's active participation in the realization of this volume, a revised edition translated into English of the existing book written in Italian.

Concerning the authors, on behalf of SIF, I would like to remember and pay tribute to Raffaella Simili, who sadly and suddenly passed away in August 2022. Moreover, I would like to express sincere thanks to Sandra Linguerri.

Their well-organized and passionate descriptions of Einstein's links with Italy will surely trigger the interest of the readers, also from a general audience, making them realize how much Einstein loved Italy.

I conclude stating that SIF is very proud to have granted its support to the realization of this editorial project.

Bologna, Italy   Angela Bracco
                                              Italian Physical Society, President

The original version of the book has been revised. A correction to this book can be found at https://doi.org/10.1007/978-3-031-52950-4_16

Preface

The plan for this book originated mainly from the results of the exhibition *Einstein in Bologna* (November 11, 2005–January 8, 2006) held as part of a network exhibition organized, under the auspices of the then Ministry of Education, University and Research together with the Universities of Bari, Pavia, and the Institute and Museum of the History of Science in Florence (now Museo Galileo) and connected with the large exhibition in Berlin, *Albert Einstein Ingenieur des Universums* curated by Jürgen Renn to celebrate the 100th anniversary of the theory of relativity.

On that occasion, we discovered how many unpublished or little known details were to be found in writings, memories, and documents of the time with regard to Einstein's visit to Bologna in October 1921 to give three public lectures at the invitation of the illustrious mathematician Federigo Enriques. Enriques was then well known for his cultural and philosophical endeavors and the documents concerned the lasting relationship of esteem and friendship that united these two great thinkers in the specialist field, a relationship that over time acquired significant elements of civil solidarity of no small importance.

Digging into Einstein's relationship with Italy, a country he knew and loved, starting from those marvelous periods in his youth spent with his family in Pavia and always nostalgically remembered by him, numerous surprising links came to light, in addition to some already known ones, with scientists, intellectuals, institutions, ministers, of the most varied nature, from the scientific/philosophical to the political. Suffice it to think of the correspondence with the mathematicians Federigo Enriques and Tullio Levi–Civita, the letters with his childhood friend Ernestina Marangoni, the political letters with Guglielmo Ferrero, Francesco Ruffini, and the philosopher Benedetto Croce, or the letter Einstein sent to the Minister of Justice Alfredo Rocco in 1931 to save his Italian university colleagues from the infamy of the oath of allegiance to the fascist regime.

Browsing through the pages of "Scientia," the international journal founded by Enriques and Eugenio Rignano in 1907, we soon noticed both the discussion on relativity promoted essentially by the editor Enriques, which included, among others, articles by Orso Mario Corbino, Guido Castelnuovo, Max Abraham, and Einstein himself; as well as the investigation, again into relativity, started in 1923 by the only remaining director, the anti-relativist Eugenio Rignano, with writings by Sir Arthur Stanley Eddington, Enrico Fermi, and Hans Reichenbach.

Here then is the purpose of disseminating this precious information to a public that is not only Italian, as was done previously (S. Linguerri, R. Simili, *Einstein parla italiano. Itinerari e Polemiche*, Pendragon, Bologna, 2008), but also internationally, so that people know how much Einstein's Italian "talking" was much richer and more apt than is usually considered and often believed abroad. In 2021, a special documentary dedicated to Albert Einstein and Italy was created by Raffaella Simili and Sandra Linguerri and produced which will be broadcast on a major Italian national television channel in the framework of the very popular and well-known show "Rai Storia."

Having reached the end of an endeavor such as the publication of a book as exciting and difficult as this one, we wish to express our deepest gratitude to President Angela Bracco and to the Honorary President of the Italian Physical Society Luisa Cifarelli for their support and encouragement. A special thanks goes to Barbara Ancarani for her fundamental help in all stages of production and editorial revision of this book. We also extend our special thanks to the archives and institutions that have enabled us to publish published and unpublished papers: the Albert Einstein Archives, the Jewish National & University Library, the Hebrew University of Jerusalem; the Burtler Library-Columbia University; the *Emilio Segrè Visual Archives*; ETH-Bibliothek Zürich, Bildarchiv; the Historical Archives and Library of the Accademia Nazionale dei Lincei e Corsiniana in Rome; the Academy of Sciences of the Institute of Bologna; the Museum for the History of the University of Pavia; the Historical Archives of the Zanichelli publishing house; the "Norberto Bobbio" Library of Turin; the Civic Museums of Pavia. Special thanks go to Federico and Lorenzo Enriques and to Enrico Sangiorgi and the Ricci-Curbastro family.

Bologna, Italy   Sandra Linguerri
July 2022   Raffaella Simili

Contents

Albert Einstein Italian Memories

Sandra Linguerri and Raffaella Simili (deceased)

On the 15th (October) I am going with my son to Bologna to give proof of my "Italian with sauerkraut".
A. Einstein to H. Anschuetz-kaempfe, 11.X.1921.

Italians will have a hard time recognizing the language of Dante in my reports.
A. Einstein to his wife Elsa, 16.X.1921.

"The Fondest Memories of Youth"

"Dear Colleague, I would be very pleased if next time you would write to me in Italian. As a young man I spent more than six months in Italy; at that time I also had the pleasure of visiting the lovely town of Padua and still today I am pleased to use my modest knowledge of the Italian language. On the other hand I would not have the courage to write to you in Italian, because the result would really be too uncertain and unclear".[1]

This is how Albert Einstein wrote from Berlin in March 1915 to the mathematician Tullio Levi–Civita, then a professor at the University of Padua, the second of those letters that gave rise to a dense exchange of correspondence destined to become famous thanks to the crucial contribution that the tensor calculus developed by Levi–Civita himself and his teacher Gregorio Ricci-Curbastro made to the final formulation of general relativity. A contribution that Einstein did not hesitate to acknowledge several times, both publicly, starting from the 1916 article[2] *Grundlage der allgemeinen Relativitätstheorie* [1] where he expounded the theory in question

---

[1] A. Einstein to T. Levi–Civita, 17 March 1915, Archivio Tullio Levi–Civita, Biblioteca dell'Accademia Nazionale dei Lincei e Corsiniana, Rome (henceforth Archivio Levi–Civita).

The first and second subsections are by Raffaella Simili, while the third and fourth subsections are by Sandra Linguerri.

[2] The article extracted from the "Annalen", was immediately published by Barth Publishing House of Leipzig and was reprinted several times. The following year, that article would become his first

in a structured manner, and privately, as can be seen from the extensive "Italian" correspondence in his archive.

Levi–Civita certainly did not begrudge the cordial request of his illustrious interlocutor and sent a "prompt" articulate reply in Italian about the mathematical problems under discussion.

"Dear Colleague, I have just received your letter of March 23 written in the Italian familiar to me of which I have been deprived for so long. You cannot imagine how much pleasure it gives me to receive a letter so authentically Italian. Reading it brings back vividly the most beautiful memories of my youth"…[3]

In short, a very nice letter—Einstein continued jokingly—perhaps in order to avoid an immediate "grumpy expression" in the face of the "implacable objections" to his calculations by the young mathematician from Padua.

It was the latter who took the initiative of writing to Einstein to ask him for some clarifications on points that were rather obscure to him. The reply was not long in coming: "Dear Colleague," Einstein wrote back on March 5, 1915, "the fact that you deal so thoroughly with my work gives me great joy. You can imagine how rare it is for someone to deal with this subject in depth with an independent and critical attitude.[4]

Here, an independent and critical attitude, combined with a thorough knowledge of the subjects he cared about, was just what Einstein needed most at that moment. A few years earlier in fact Einstein, trapped in the intricate labyrinth of gravitational equations among which he had not yet found the right solution, had launched the well-known cry for help "Grossmann, you must help me or else I'll go crazy" ([4], p. 212) to his friend Marcel Grossmann, a mathematician at the Zurich Polytechnic and old fellow-student to whom Einstein had dedicated his doctoral thesis.

It was in fact Grossmann who put his hands on the possible way out represented by the absolute differential calculus of Ricci-Curbastro and Levi–Civita.[5] It was indeed the common objective of Grossmann and Einstein, who published together some papers between 1913 and 1914, to try to place this calculus at the foundation of the general theory of relativity. It was precisely on the basis of these publications, and on others written by Einstein alone, that Levi–Civita's "implacable objections" intervened in the cordial "Italian" letter.

Their correspondence, though more sporadic, continued for a long time until 1930 and marked a collaboration not only in the scientific field but also a closeness on the human level.

It was in 1905, his *annus mirabilis*, that Einstein, while at the Patent Office in Bern, published four memorable articles destined to cause a revolution in physics

---

book and also his most widely known and most translated work [2]. The first Italian translation is dated 1921 [3].

[3] A. Einstein to T. Levi–Civita, 26 March 1915, Archivio Levi–Civita.

[4] A. Einstein to T. Levi–Civita, March 5, 1915, Archivio Levi–Civita.

[5] This calculus, disclosed in 1900 in the paper *Méthodes de calcul differentiel absolu et leur applications* [5], had been elaborated in its essential points by Gregorio Ricci-Curbastro well before the *Méthodes* and developed by Levi–Civita, who then framed it in new geometrical visions.

and in science in general, but which at the time aroused little interest if not general skepticism, with the exception of the eminent German physicist Max Planck who, as editor of the "Annalen der Physik", had favored their publication. He later became Einstein's privileged interlocutor and convinced supporter. In fact, it was he who in 1913 invited Einstein to move to Berlin where he settled in 1914 and from where he exchanged letters with Levi–Civita.

In the first article entitled *Über einen die Erzeugung und Verwandlung des Lichtes betreffenden heuristischen Gesichtspunkt* [6], Einstein was able to demonstrate that the quantum theory proposed in 1900 by Planck could be used for the physical explanation of many difficult to understand phenomena related to electromagnetic radiation. With the second, *Die von der molekularkinetischen Theorie der Wärme gefordete Bewegung von in ruhenden Flüssigkeiten suspendierten Teilchen* [7], he contributed decisively to the determination of the existence of atoms, thus providing, on the basis of Brownian motion, significant support for Boltzmann's theories.

It was in the third, *Elektrodynamik bewegter Körper* [8], that Einstein revised the traditional concepts of space and time in an attempt to resolve the contrasts between Newton's and Maxwell's theories. In order to unify them Einstein elaborated restricted or special relativity (i.e. limited to a principle of relativity related only to uniform rectilinear motions).

With the theory of relativity, space and time found themselves inextricably linked. As we approach speeds close to that of light time dilates (i.e. it flows more slowly), while distances to be traveled become shorter.

In volume XVII of the "Annalen", also in 1905, Einstein published a fourth article *Ist die Trägheit eines Körpers von seinem Energienhalt abhängig?* [9] in which he stated the original formulation of the famous equation $E = mc^2$.

With regard to the four papers Einstein told Conrad Habicht, a mathematician and fellow-student in Bern, that the first one was extremely innovative. At the end of the one on special relativity he thanked Michele Besso, a Swiss-Italian engineer who at the time was his colleague at the Patent Office, but above all one of his two "friends for life" (the other was Heinrich Zangger) whom he had met in 1895: "lastly I wish to say that in working on the problem treated here I had the loyal support of my friend and colleague Besso and that I am grateful to him for various valuable suggestions" ([9], pp. 94–95).

Perhaps for this reason in 1907 Besso presented his essay on relativity to the mathematician Federigo Enriques [10, 11], then editor of the international periodical "Rivista di scienza", who seems to have had some reservations, so much so that neither in that year nor in the following ones did anything by Besso appear on this subject [12].

By that time probably Enriques and Besso had met as early as 1890 in Rome where the latter was a student in the first year of university at the Faculty of Science. Michele's visits to his aunt and uncle who lived in the capital were frequent. Moreover his uncle Davide, an authoritative professor of mathematics at the Technical Institute of Rome and from 1888 of algebra at the University of Modena, was the director of the "Periodico di Matematiche" from 1886 to 1896, a periodical that Enriques himself edited from 1921 to 1938 [13–15].

What is certain is that, in a letter sent in 1917 to Einstein by another "friend for life" Zangger [16], an important personality at the University of Zurich and director of the Institute of Forensic Medicine, who had studied in Naples and Zurich then specializing in Paris, forwarded to him the greetings of Besso and Enriques who were with him in the city at that time.[6]

In the years following the formulation of special relativity, Einstein, as is known, continued his research intensively, expanding his ideas on quanta, Brownian motion, and gradually registering a growing interest in relativity.

Academically, he was appointed first at the University of Zurich and then, in 1911, at the German University of Prague.[7] On the scientific side, the springboard was an invitation to the Congress of Scientists held in Salzburg in 1909, where he presented a paper on relativity and quantum particles. However, it was his participation in the Solvay Congress—named after its financial backer—organized in Brussels by the German physicist Walter Nernst, that established his international authority. There were a few select invited participants, among them Marie Curie, Paul Langevin, Jean Perrin, Henri Poincaré, Arnold Sommerfeld, Ernest Rutherford, James Jeans, to name a few. Their meetings were chaired by the Dutch physicist Hendrik A. Lorentz.

After a brief return to the Zurich Polytechnic in 1912, Einstein became a professor in Berlin in 1914, the director of a newly established research institute, free from teaching duties, and an influential member of the Prussian Academy of Sciences. Everything went smoothly, including the condition he set for the transfer, namely to remain a Swiss citizen[8] (Jewish) and not to become a German citizen again. However events had gone less smoothly in his family since, at that time, Albert separated from his wife Mileva Marić [19] who returned to Switzerland with their sons Hans Albert and Eduard. Einstein then again married to his cousin, Elsa Löwenthal, who already had two daughters, Ilse and Margot.

In Berlin, in addition to carrying out the final formulation of general relativity, he became aware, thanks to the Zionist movement, of his Jewish identity as well as of the difficulties of his "people".

It was on the subject of the project to establish a university in Jerusalem that Einstein came into contact with Chaim Weizmann, the undisputed leader of that movement and future president of the State of Israel.

A professor of chemistry at the University of Manchester, his wartime scientific work had rendered great services to the British government during the world war. He had thus come into contact with influential circles that helped him propagate Zionist ideas. In 1921, before visiting Bologna, Einstein and Weizmann went to the United States to raise funds for the future Hebrew University, convinced that it could serve as a valuable link between East and West.

---

[6] H. Zangger to A. Einstein, Zurich 17 December 1917 [17].

[7] On his stay in Prague, see, in particular, [18].

[8] After renouncing his German citizenship in 1896 he became stateless to then become a Swiss citizen in 1901. In 1940 he asked for and was given US citizenship while keeping the Swiss one as well.

Following the publication of his 1916 paper *Grundlage der allgemeinen Relativitätstheorie*, in which he explicitly acknowledged the merits of Ricci and Levi–Civita's algorithmic "system" applied to problems of theoretical physics, Italy, no differently from other European countries and overseas, played its part in opening a lively debate in the scientific community, a debate whose content was rather heterogeneous and which, to tell the truth, had been going on for some time, albeit more quietly. If, on the one hand, it had aroused a positive interest among the mathematicians Guido Castelnuovo, Enriques, Levi–Civita, on the other hand it had met with strong criticism among physicists and astronomers, criticism that gradually took on the form of outright controversies.

These controversies, however, did not leave any negative mark on Einstein who, on the contrary, always had words of esteem and sympathy for the Italian scientific community.

In fact, when in our country it was decided to celebrate the fiftieth anniversary of relativity with a volume, *Cinquant'anni di relatività 1905–1955*, edited by Mario Pantaleo, Einstein included in the preface the following sentence: "this book was born from the friendly feelings of *my* Italian colleagues" [20].

On the other hand, the great scientist's ties with Italy were many and long-lasting, starting from those "marvelous and vivid memories" of a youth happily spent in Pavia that Einstein mentioned in his letters to Levi–Civita,[9] memories that he kept all his life, according to the testimony of his faithful secretary, Helen Dukas.[10]

It was during an interview on October 22, 1921, given by Einstein to the correspondent of the "Corriere della Sera" on the occasion of the three public lectures held in Bologna at the Archiginnasio (the ancient seat of Bologna University) at the invitation of Federigo Enriques, that he himself, in talking about his life and work, nostalgically recalled "certain memories of his youth"—the journalist pointed out—having "lived as a child for six months in Lombardy before settling in Switzerland; Indeed, his father, who was an electrician (*sic*) had lived first in Pavia, then in Milan for many years. It is therefore with a sympathy not lacking a personal note—the article ended—that Einstein speaks of Italy".[11]

Now, Einstein's father was not an electrician at all, but an owner of electrical factories who settled, with his engineer brother Jacob and their respective families due to financial difficulties first in Pavia then in Milan. In fact, after a few years of prosperity, Herman and Jacob were forced to liquidate their small factory in Munich and turn their attention to the city of Pavia, where they set up a new factory in partnership with an Italian engineer, Lorenzo Garrone.

---

[9] For further visits to his family in Milan see [21].

[10] During the memorable year of imaginative freedom in Italy Einstein "with his friends Otto Neustätter hiked fancy-free through the Apennine mountains to Genoa, where he had relatives. Museums, art treasures, churches, concerts, books and more books, family, friends, the warm Italian sun, the free, warmhearted people-all merged into a heady adventure of escape and wonderfull self-discovery" [22].

[11] *Come vive e lavora Einstein l'inventore della teoria sulla relatività,* "Corriere della Sera", 22 October 1921.

Albert, however, remained in Munich to complete his studies until, in the grip of the strongest discomfort caused, according to him, by the over restrictive and militaristic mentality of the school, he took refuge with his family in Pavia. Young Albert must have been convincing since he did not receive any reproach from his father, on the contrary he convinced him to "free" him from German citizenship thus avoiding the obligation of military service and to send him to continue his studies in Switzerland, as in fact happened, at the school of Aarau (where he met Habicht and Maurice Solovine with whom he established a fraternal friendship and shared his first cultural interests) and later at the Polytechnic in Zurich.

It was in fact his sister Maja who was the first to give news, in a biography in 1924 dedicated to her brother [23], of his stay in Pavia, of the family's long series of reversals that continued, alas, even after this stay and, above all, of the carefree life spent by her and Albert in the country house, in Casteggio, with her friend Ernestina Marangoni.

The "most marvelous memories" of the happy youth in Pavia were splendidly recalled many years later, to be precise on May 14 1955, less than a month after Einstein's death, by Ernestina in person, who also made known the correspondence between them in 1946, after the war was over.

"I met Albert on a summer afternoon when he was on holiday between two semesters, at the bathing establishment in Ticino, and he seemed to me a rather delicate but healthy young man—a little wan in color, dark eyes, brown hair not black like his sister's and as it became later". In these summer encounters Albert, who "spoke Italian quite well—Ernestina continued—cast his *boutades* into German, leaving them to be translated by his sister".

"The happy months of my stay in Italy—Einstein had written to her in Italian, signing himself Alberto in the first of his letters in 1946—are the most marvelous memories [...] Days and weeks without anxiety and without tension [...]".[12]

"The most marvelous days—a moved Ernestina echoed Albert's words—were spent on the hills for festive harvests in the vineyards rich with grapes [...] in a group of close-knit and homogeneous youth; or listening to Albert accompanied on the piano by his sister Maja, playing with passion an excellent violin that he hadn't brought with him but that I borrowed for him" [24, 26].

Finally, Ernestina added to these chronicles other episodes, including scientific ones, concerning the young Einstein, while amusedly describing the brilliant trip he made with his friend Otto Neustätter via tram from Pavia and on foot from Voghera to visit an uncle in Genoa.

From his sister Maja we learn of another Italian friendship of Albert's with a politically influential personage he met on holiday in Airolo on the Gotthard Pass, Luigi Luzzatti, a minister several times in various governments and even Prime Minister.

---

[12] A. Einstein to E. Marangoni, 16 August 1946, Museo per la Storia dell'Università di Pavia, in [25].

In 1896, Einstein, as decided in the family, left for Switzerland to resume his studies, thus interrupting his carefree Italian season and beginning his scientific career which was certainly not easy but extraordinarily successful.

Maja and Ernestina remained in contact for a long time. Their friendship was fostered by the fact that Maja, married to Paul Winteler, after a stay in Lucerne, lived in Colonnata near Florence until 1939. It was precisely on this date that the two friends met for the last time in Milan. Following the racial laws Maja fled to take refuge in the United States where her brother Albert, who in turn had fled Berlin immediately after the Nazi takeover, was already living.

"I have been living in America [...] lacking a real trust in men and God—Einstein vented anguishedly to Ernestina—a wandering life with a unique constant mathematical work [...] I am glad to hear that all the Casteggio friends [...] are safe [...] dear Mussolini as honestly deserved".[13]

In a second letter sent to "dear Ernestina", also written in English ("here in America, my Italian has become too rusty"), Einstein resumed talking about his loneliness: "It is a strange thing to be so widely known and yet to be so lonely. But it is a fact that this kind of popularity—as it has become the case with me—is forcing its victim into a defensive position which leads to isolation". And he concluded: "what beautiful memory is Casteggio. What charm has the little town seen with the admiring eyes of youth".[14]

But gradually the memories faded and the sad reflections resumed. "We had to witness gigantic political upheavals and shall witness still more if we are not called away in time. Essentially everything is always the same. The nations always walk again into the trap, because the atavistic impulses are stronger than reason and acquired *[sic]* convictions".[15]

Also after Einstein's death and a few days before Ernestina's, another elderly lady opened the treasure chest of her memories as a young girl: Adriana Enriques, the mathematician's daughter, who had met Einstein in Bologna in 1921. And what a meeting! Not only did she have the opportunity to get to know him well privately, at her family's home, during social receptions and scientifically intense discussions with her father, Levi–Civita and others, to guide him through the streets and monuments of the city, to attend his lectures, but also to attend a lecture specially prepared by Einstein for students.

"We young people enjoyed the privilege of a special meeting (from which the professors were excluded), in which the scientist answered all our questions with cordiality and clarity. I remember that the future professors Chisini, Todesco and Notari were present" [27].

In fact, Adriana was "reluctant to go" at first; she considered herself "still too ignorant of these things. "Come and you will understand," Einstein insisted. "I went

---

[13] A. Einstein to E. Marangoni, 16 August 1946, Museo per la Storia dell'Università di Pavia, in [25].

[14] A. Einstein to E. Marangoni, 1 October 1952, Museo per la Storia dell'Università di Pavia, in [25].

[15] See footnote 14.

and I understood; he was very clear and used simple words. He explained the fourth dimension with the journey of a sole that, being flat, and living in an apparently flat space, is unable to suspect that the Earth is spherical: it lacks the intuition of which those who resign themselves to the awareness of the three dimensions are deprived" [28, 29].[16]

At the time of his departure, in a small album carefully preserved by Adriana over the years, Einstein wrote by hand an aphorism dedicated to her in memory of the acquaintance he had made in October 1921: "Study and, more generally, the love of beauty and truth, are things before which one would always want to remain a child. To Adriana in memory of the acquaintance made in October 1921. Albert Einstein" [30].

A week before his death, Einstein had time to sign the last of the many manifestos carried out in his pacifist campaigns, the manifesto promoted by Bertrand Russell against the use of the atomic bomb. The text read: "We appeal as human beings to human beings: Remember your humanity, and forget the rest. If you can do so, the way lies open to a new Paradise; if you cannot, there lies before you the risk of universal death".[17]

Einstein's support was courageous, given the political climate of witch-hunting then prevailing in the United States, in the midst of the Cold War with the USSR, and seeking super-bombs. A climate that, as is well known, favored the nefarious activities of Senator Joseph McCarthy, frequently condemned by Einstein himself in the press, as well as the trial personally led by the ex-general president Eisenhower against his former collaborator the physicist Robert Oppenheimer, who had directed the Los Alamos laboratories for the construction of the atomic bomb during the world war.

The Manifesto against the atomic bomb, in Einstein's case, drew its origins far back in time from the *Manifesto to Europeans for Peace and Unity of Peoples* presented at the outbreak of World War I [31] by the German physiologist Georg Friedrich Nicolai, a well-known pacifist, by the former director of the Berlin Observatory, Wilhelm Föster, by Otto Buek of Heidelberg and by Einstein himself.

The pacifist manifesto of these four men was intended to be a response against war and nationalism in the name of which, in defense of their homeland, no less than ninety-three highly prestigious intellectuals –including Felix Klein, Max Planck, Wilhelm Ostwald, Walter Nernst, William Röntgen, Wilhem Wundt, and Wilhem Windelband – had signed an appeal addressed to the "civilized nations" in October 1914. The appeal rejected any German responsibility for the conflict, defended the invasion of Belgium, denied the alleged atrocities, and asserted that German culture and militarism were one and the same ([32], pp. 4–7).

The four signatories remained completely isolated despite the fact that the document circulated widely in university and cultural circles in general: "Our sole purpose

---

[16] A. Enriques to O. Nathan, Turin 17 February 1957.

[17] The Russell/Einstein Manifesto was the trigger for the creation of what would later become the Pugwash movement of the scientists against the atomic bomb, that was awarded the Nobel Peace Price in 1995.

is to affirm our profound conviction that the time has come when Europe must unite to guard its soil, its people, and its culture. We are stating publicly our faith in European unity, a faith which we believe is shared by many; we hope that this public affirmation of our faith may contribute to the growth of a powerful movement toward such unity. The first step in this direction would be for all those who truly cherish the culture of Europe to join forces - all those whom Goethe once prophetically called "good Europeans". We must not abandon hope that their voice speaking in unison may even today rise above the clash of arms..." ([32], pp. 5, 6).

This pacifism was unknown in Germany and certainly did not favor Einstein, who had just arrived in Berlin with a Swiss citizenship in his pocket, a pacifism that only aroused mistrust and suspicion towards him. Unperturbed, he tried to get in touch with pacifists from other countries, including the intellectual Romain Rolland, whom he visited in Geneva in September 1915, thus marking another step on his uninterrupted path towards world peace.

During the war, in spite of the din of arms, he worked intensively, publishing, as we have said, a series of memoirs on general relativity, as well as a booklet that explained relativity in popular terms, a booklet *Über die Spezielle und die Allgemeine Relativitätstheorie, Gemeinverständlich,* which was so successful that in 1918 a third reprint was even considered [2].

However, it was with the experimental confirmation achieved after several unsuccessful attempts during the famous expeditions on solar eclipses thanks to the astronomers Arthur Eddington and Frank Dyson in Africa and Brazil, to which we will return later, that the Einstein legend exploded.

The results were presented in London at a joint meeting of the Royal Society and the Royal Astronomical Society on November 6, 1919. The then president of the Royal Society, Joseph J. Thomson, declared that it was the most extraordinary scientific event since the discovery in 1846 of the planet Neptune. This session was "historic", according to the words of the philosopher and mathematician Alfred North Whitehead: "It was my good fortune to be present at the meeting of the Royal Society in London when the Astronomer Royal for England announced that the photographic plates of the famous eclipse, as measured by his colleagues in Greenwich Observatory, had verified the prediction of Einstein that rays of light are bent as they pass in the neighborhood of the sun. The whole atmosphere of tense interest was exactly that of the Greek drama: we were the chorus commenting on the decree of destiny as disclosed in the development of a supreme incident. There was dramatic quality in the very staging:—the traditional ceremonial, and in the background the picture of Newton to remind us that the greatest of scientific generalizations was now, after more than two centuries, to receive its first modification" [33, 34].

In the aftermath of the "historic" session the English newspapers exploded with enthusiastic comments talking about *a revolution in science, announcing a new theory of the Universe and overturning Newton's ideas.*

On November 9, the "New York Times" reported the news in an article entitled, All the *lights in the sky are crooked,* a piece of news to which it returned several times later in more and more exalted tones. On December 14, the magazine "Berliner Illustrirte Zeitung" published a cover story with a photograph of the "new giant of world history", Albert Einstein. Other newspapers followed suit: From Europe to America, Einstein's theory triumphed.

In Italy it was the "Corriere della Sera" that gave the news of the event with the following words: "the prophecy of a scientist". Thus began Einstein's worldwide fame.

In the face of this triumph this charismatic man never lost his head. On the contrary, when he went to England for a series of lectures, he replied with biting self-mockery to the pressing questions of the journalists by stating that the theory of relativity was subject to the tastes of the readers. While in Germany he was considered a German scientist and in England a Swiss Jew, in the future, if he became an archenemy, it would be the other way around: Swiss Jew for the Germans and German scientist for the English [35].

Barely a year later, in February 1920, the "hero" Einstein was subjected to an anti-Semitic protest at a conference in Berlin. A few months later there was another demonstration, this time well organized by nationalist movements, against the theory of general relativity. And this was only the beginning. Shortly thereafter, Einstein's friend, the (Jewish) foreign minister Walther Rathenau, was assassinated.

In 1920 the League of Nations was founded, a supranational institution created to promote peace among peoples, from which however Germany was excluded.

In 1922, Einstein, Marie Curie, Henri Bergson, Gilbert Murray were invited to join an important agency of the Society, the International Committee for Intellectual Cooperation ([32] pp. 58–89).

Einstein enthusiastically accepted in the hope of serving the international cause of peace. Later that year he was awarded the Nobel Prize in Physics for 1921 [36].

His proposal for the Committee, as testified by the Oxford classicist Murray, who later collaborated with Einstein in opposing the Italian Fascist oath of 1931, provoked negative reactions both from the Germans (he was still a Swiss Jew, although famous) and the French (he was a German). But it was not these reactions that upset him, but rather the perplexity and disappointment that he was gradually accumulating with regard to the work of the Committee, which in his opinion was not sufficiently impartial. For this reason he announced his resignation in March 1923 and then withdrew it in 1924.

It was in this Committee that he made the acquaintance of the jurist Francesco Ruffini, professor at the University of Turin, senator from 1914, Minister of Education in 1916–17, known for his frequent speeches in the Senate together with those of his colleagues Croce and Volterra against the Mussolini government.

Senator Ruffini paid dearly for this opposition, starting with his resignation from the Committee, which he was forced to do in 1925 to make way for the Fascist Minister of Justice, Alfredo Rocco.

Both Marie Curie and Einstein took up the cudgels against Rocco's appointment. Apart from their solidarity with their colleague Ruffini, they categorically rejected appointments of a political nature, especially if they came from totalitarian regimes. Einstein himself wrote about Rocco to his wife Elsa, on February 17, 1926: "[...] Yesterday morning I had a furious battle with the Fascist member of the committee. He will remember it as long as he lives" ([32], p. 78).

Einstein was so enraged against the representative of the Italian Fascist regime that he went so far, contrary to his custom, as to put himself forward in place of Rocco on the board of the Committee. He resigned only when Italy threatened to leave the League of Nations because of this incident.

Einstein's membership of the Committee became increasingly difficult due to the fierce political attacks on him from Germany, until he resigned in 1932.

A year later, with the Nazi party in power, the persecution of him became official: at that time he was on a study trip in the United States and solemnly announced that he would never return to the country of his birth with a cry of accusation: "I am not going home!" ([32], p. 211). And he never returned, spending the rest of his long scientific and political life in Princeton in the United States at the Institute for Advanced Study where he continued to pursue his research.

Before being expelled, however, he had time to resign from the Prussian and Bavarian Academies. This was a stratagem he also used in 1938 when the Accademia dei Lincei, of which he had been a member since 1921, was preparing to expel its Jewish members following the racial laws.[18] Many of "his" Italian colleagues were expelled with him: he was reinstated like all of them with a decree of April 12, 1945.

In 1946, the year that marked the "rebirth" of the Academy after the terrible catastrophe of the dictatorship and the Second World War, Einstein took up his pen to express to Guido Castelnuovo, who had also just been reinstated, newly appointed president of the Accademia dei Lincei and later senator for life, his great pleasure that the Academy "has resumed its activities for the benefit of science, your country having been liberated from fascist oppression", ending with a nice "affectionate greeting" in Italian.[19]

Enriques, too, managed to regain his position at the University, but not for long; he died in 1946. Levi–Civita, on the other hand, did not make it; he passed away after a long illness in 1941, a year after the death of the great mathematician Vito Volterra, known in the world as "Mr. Italian Science" [37–39].

Levi–Civita had time, however, a few years earlier, to bear a final testimony of his esteem and friendship to his famous friend Einstein. In 1936, in fact, he was contacted by the president of the Pontificia Academia Scientiarum and rector of the Catholic University of Milan, Agostino Gemelli, to propose new appointments of Italian and foreign members.

Of the seventy academics, thirty had to be Italian and belong to the following groups: physics, astronomy, mathematics and applied sciences, non biological sciences (chemistry, geology, mineralogy, geography), biological sciences, while "for the foreigners – Gemelli specified—we must choose men of the first order and beyond question, the choice can also fall on non-Catholic men, as long as they have not written or published or said anything against Catholicism in particular and against Religion

---

[18] A. Einstein to Reale Accademia Nazionale dei Lincei, 15 December 1938, Historical Archive, Accademia Nazionale dei Lincei, Rome.

[19] A. Einstein to G. Castelnuovo, 26 June 1946, Historical Archive, Accademia Nazionale dei Lincei, Rome.

in general".[20] Of course, among the theoretical physicists, Levi–Civita immediately recommended Einstein's candidature, characterizing him as follows: "builder of the theory of relativity, which is a synthesis of natural laws, more complete and refined than Newton's: correcting, in a quantitatively light, conceptually profound way, the principles of mechanics and physics, he predicted new, delicate astronomical and spectroscopic phenomena that were fully confirmed experimentally; Nobel Prize".[21]

In a previous letter, he had had no hesitation in mentioning, correctly following Gemelli's indications, that Einstein was a "celebrated non-Catholic".[22]

When Levi–Civita put forward his proposals for the Italians, he gave "a place apart to three people",[23] namely Volterra, Enriques and Fermi, specifying that both Volterra and Enriques were Jewish. Of the three "separate places" requested by Levi–Civita, only one, that for Volterra, was accepted. On the other hand, there were many gray names added and/or substituted by the President of the Pontifical Academy and then by the Pope, despite Levi–Civita's lively protests.

Among the foreign members, among those men "of the first order and beyond question" invoked by Father Gemelli to give prestige to his Academy, the name of Einstein did not appear. The proposal was not accepted.

In Bologna and After: From Science to Politics

"Dear Michele we are here [with his sister Maja] in Florence—Einstein wrote in a postcard to his friend Besso on October 20, 1921—tomorrow I am leaving for Bologna where I have to give a lecture in Italian, poor things!".[24]

It was his visit to Bologna, the only Italian city where Einstein gave three lectures on relativity open to the public. Einstein came there at the invitation of Federigo Enriques, the illustrious mathematician who was then a professor in Bologna, an enterprising organizer of cultural events and former editor of the journal "Scientia" [41–44].

Einstein and Enriques had written at length about these conferences. It seems, among other things, that their contacts dated well before the letter of greetings written by Zangger in 1917, which we mentioned in the previously.[25]

---

[20] A. Gemelli to T. Levi–Civita, Milan 20 January 1936, Archivio Levi–Civita.

[21] T. Levi–Civita to A. Gemelli, Rome 27 January 1936, Archivio Levi–Civita.

[22] T. Levi–Civita to A. Gemelli, Rome 2 January 1936, Archivio Levi–Civita.

[23] T. Levi–Civita to A. Gemelli, Rome 27 January 1936, Archivio Levi–Civita.

[24] A. Einstein to M. Besso, Florence 20 October 1921, in [40].

[25] In 1913 Enriques was among the promoters of a conference of mathematicians and philosophers, planned by the French Philosophical Society, to be held in Paris in 1914 where Einstein himself had been invited. However, the outbreak of war prevented the conference from being held, despite the preparatory meetings. See [45, 46]. Enriques had, however, frequently stayed in Zurich, where both Zangger and Besso lived, starting in 1897 on the occasion of the First International Congress of Mathematicians and especially in 1914 at the bedside of his wife, who had undergone eye surgery.

In any case, in 1920, Enriques apologized for not having been able to reply to a postcard from Einstein due to "the unfortunate war" and hoped to be forgiven by expressing "consent and esteem" to "a superior spirit like his".[26] In continuing, Enriques, after a few pleasantries, expressed, on the one hand, his most fervent admiration for Einstein's latest great success, general relativity; on the other hand, he confessed to having "not assimilated well the spirit of his guiding ideas", wishing to meet him "in conditions favorable to a relaxed conversation".

The occasion to satisfy this "lively aspiration" presented itself not even a year later: in a letter dated 19 January 1921, Enriques informed Einstein that it had been decided to have him inaugurate ("No name has seemed equal to his and no subject so exciting for the scientific world as relativity") a series of meetings promoted by the University to invite renowned foreign intellectuals to our country.

Einstein replied in the affirmative, nostalgically recalling his "marvelous memories of his youth" [27]. Enriques was also enthusiastic about the fact that the lectures would be in Italian; "which would make his words even more fruitful".[27]

At the end of September Enriques moved on to the organizational phase: "Dear and illustrious colleague, I receive with great interest your kind letter, from which I have the confirmation of your coming to Bologna, in the second half of next month. Consequently, your three lectures remain fixed for Saturday 22, Monday 24 and Wednesday 26 October. In this way, as you kindly hint, [...] there will be an opportunity to arrange some meetings in *petit comité*, since many of us are anxious to have many explanations from you, as you can well imagine [...]. In any case, I beg you to come to my house, for tea, on the evening of Friday 21 at 9 o'clock (that is 9 p.m.) [...]. It is understood that your son, who is just the same age as mine, will come with you".[28]

Thus it was that Albert Einstein held three lectures on 22, 24 and 26 October 1921 at the Archiginnasio, home of the Antico Studio, which were a great success with the public.

Of those conferences, two memorable photographs remain, one portraying Albert Einstein and Federigo Enriques in the loggias of the Archiginnasio; the other, Einstein, Enriques and a group of professors of the University, both taken, it seems, by Enriques' young son, Giovanni, with a Kodak[29] camera.

Later, he went to the University of Padua to pay homage to the creator of absolute differential calculus, Gregorio Ricci-Curbastro[30] and to personally show him all his esteem and deepest gratitude. Albert Einstein—so it is told—"spoke for an hour, slowly, accurately, in Italian, with a scientific precision that seemed almost to acquire

---

[26] F. Enriques to A. Einstein, 20 April 1920, The Albert Einstein Archives, The Jewish National & University Library, The Hebrew University of Jerusalem (henceforth Einstein Archives).

[27] F. Enriques to A. Einstein, Bologna 17 February 1921, Einstein Archives.

[28] F. Enriques to A. Einstein, Bologna 25 September 1921 Einstein Archives.

[29] Recollection of Andrea De Benedetti, Adriana's son, communicated during a telephone conversation during the preparation of the exhibition "Einstein a Bologna " (Bologna, 11 November 2005–8 January 2006).

[30] "Il Veneto", 28–29 October 1921, p. 5. Among the numerous recollections of Ricci-Curbastro's work, see [47]. On the figure of Ricci-Curbastro see [48, 49, 50, 51].

prominence and perfection from the linguistic precision of the speaker, a good master, absolute and gifted, of our language".

It fell to Enriques' second daughter, Adriana, the other "lady of memories," then a freshman mathematics student, to welcome the Einsteins father and son upon their arrival in Bologna.

"Under the platform roof of Bologna station Anselmo Turazza, Emilio Supino and I, in our student caps, were waiting for the train to arrive. How will we recognize him? Will he travel first or second class? Will he speak Italian? It seemed to us modest freshmen at the University that the task of receiving Einstein was much more than we deserved" [27].

But then—she continued her story - "a prolonged whistle was heard: the train was entering the station. Emilio had to watch the second-class travelers, Anselmo the first-class ones. I wandered here and there, lost and trembling. But when a tall, imposing-looking gentleman stepped out of a third-class carriage, wearing a wide-brimmed black hat like those worn by artists, his hair falling down to his ears, we had no doubts. All three of us, making our way through the crowd that filled the station, rushed towards this gentleman. It was him, it could only be him, Albert Einstein. We didn't even know him from photographs, yet we would have recognized him among thousands of travelers. The imprint of genius seemed to be written on his brow" ([27], p. 21).

At the *soirée* on the 21st organized at Enriques' house in honor of the great scientist, some journalists were also present: "Prof. Alberto Einstein from Florence arrived in Bologna today", wrote the correspondent of the "Corriere della Sera", telling about the reception offered at the house of Prof. Enriques, who had gathered around his illustrious guest "the cream of Bolognese intellectuals... a few minutes after his arrival, he was already engulfed in a discussion with two luminaries of Italian science, Levi–Civita and Majorana, one a great mathematician, the other a great physicist".[31]

Reports of his lectures were widely reported in major local and national newspapers.

According to their accounts, Einstein's impact on the audience was as theatrical as "that of a bel canto star on an opera stage". At the first lecture, the curiosity among all those who on every occasion want to be able to say "I've been there too"[32] was so great that the Stabat Mater, the Aula Magna of the Archiginnasio, was completely inadequate. Even in the following days, the interest in the lectures on relativity continued to increase, attracting the usual "large audience", especially of students, who had become accustomed to scattering in the hallway of the Archiginnasio and under the portico of the Pavaglione.[33]

According to Adriana Enriques, "almost no one understood the theory of relativity, yet the Archiginnasio was packed not only with scientists from all over Italy, but also

---

[31] *Come vive e lavora Einstein l'inventore della teoria della relatività*, see footnote 11.

[32] *La prima conferenza di Einstein*, "Corriere della Sera", 23 October 1921.

[33] *La seconda conferenza di Einstein. La teoria della relatività generale*, "Corriere della Sera", 25 October 1921.

with students from all faculties, humble artisans and workers. The people were moved
as Einstein passed by and followed him in long columns, as if to show their gratitude
for having chosen Bologna for his first contact with Italy after so many years of
absence" [27].

"Albert Einstein does not need to be presented to you"—so began Enriques
pompously at the beginning of his speech of introduction, according to a skillful
presentation which he was by now an expert in, having successfully put it to the test
at the International Congress of Philosophy in 1911 [52, 53]—"the fame that is, in
this case, the herald of glory, has already shouted his name" [54].

His theories, "among the most abstruse [...] today attract the attention of the
whole world, and not only of mathematicians or philosophers, but also of the general
public"—Enriques continued—"he subverts with his criticism the common concepts
of space, time and movement [...] reason does not have necessarily marked limits,
[...] it does not offer experimental data a pre-established order, but finds in itself the
power to enlarge the frameworks in which the familiar experience of nearby things
is composed [...]. Certainly there is something surprising and almost frightening
in this progress of thought that goes beyond the limits of one's own intuition [...]"
([54], pp. 329–330).

His "revolutionary" doctrine has offered "a new occasion to call out the bankruptcy
of science"—underlined Enriques, reiterating polemical positions already taken in
the past—but "whoever judges in such a way is far away, not only from Einstein's
thought, but also from the historical concept of science [...]". "So Einstein's theory
does not mean the death of Newton's theory but on the contrary, the conquest of
a truer truth [...]". This is—Enriques concluded in the name of that banner made
of reason and freedom he always practiced during his intellectual career – "the real
reason for the emotion aroused by Albert Einstein. He restores our faith in reason"
or rather "invites us" to "turn to the contemplation of the order that the mind is able
to discover outside itself, in the marvelous work of art of nature". This is why it
was necessary to listen "reverently to the word of the Master: with that reverence
which is due to free spirits and which is meant to be a free discussion of ideas" ([54],
pp. 330–331).

Taking the floor in Italian, the Master explained first of all how the theory of relativity had arisen from problems deriving directly or indirectly from experience such
as the concept of contemporaneity taken to be self-evident as well as the presumption
deriving from habit that consists in believing that the form of bodies is independent
of their movement.

He then went on to illustrate the theory of special relativity starting from the
principle of relativity in the strict sense and from the law of the constant speed of
propagation of light in different directions, independently of the motion of the light
source, in order to show how these two valid laws, derived from experiment, are only
apparently in conflict with each other.

In the second lecture, while he briefly summarized the concepts of special relativity expounded in the first lecture, he tackled the problems connected to general
relativity pointing out, among other things, how methods of measure furnished by
non Euclidean geometry were connected to it and how it was possible to extend to

accelerated (or non inertial) systems the results of the special theory of relativity and to arrive at knowledge of the general law of the gravitational field. This law is not a new law with respect to Newton's classical law, but a broader one, since it accounts for certain anomalies related to the planets that Newton's law alone cannot explain.

In the third lecture he dwelt on a consequence of the theory, susceptible to experimental confirmation, to then go on to expound, on the basis of all the obtained results, the "relativistic" concept of the universe.

At the end of this lecture, Einstein did not fail to add that the mathematical tool he had used was derived from the calculations of Gauss and Riemann as well as Ricci and Levi–Civita.

It was the latter who was interviewed at the end of the lectures by the correspondent of the "Messaggero". In this article, too, the journalist naturally mentioned the revolutionary character of the theory as well as the Italian contribution of Ricci and Levi–Civita, and out of patriotism he also mentioned Guido Castelnuovo, Roberto Marcolongo, Gian Antonio Maggi and Attilio Palatini.

Levi–Civita, who was first asked by the journalist to outline the characteristic features of relativity, had the opportunity to clarify that relativity was certainly not "a subject for a political journal, nor one that could be easily popularized". Moreover, it was necessary "to begin by distinguishing between the revolution in the mathematical representation of physical concepts and philosophical speculation"; in this sense "one can be neither a relativist nor an anti-relativist".[34]

In this regard, Levi–Civita was under an illusion, since the very occasion of Einstein's visit to Bologna fostered a series of disconcerting reactions not only in scientific circles but also in journalistic circles and in the more eccentric cultural circles.

Returning to the days in Bologna, that great scientist Einstein also bid farewell to the city with a meeting at the Academy of Sciences, where an extraordinary session was held for the purpose of studying in depth some points of the theory of relativity.

The president, Giuseppe Ruggi, extended a greeting of welcome to Einstein, who had been elected a foreign member in April 1921,[35] while Enriques outlined the lines of discussion[36].

The speakers were Burgatti, Ciamician, Majorana, Amaduzzi, Levi–Civita, Castelnuovo and Enriques himself. According to the minutes, Einstein replied brilliantly to all of them, thus earning the president's warmest thanks.

A few months later, Enriques, who had been the scientific consultant of Zanichelli since 1906, furthered the first Italian translation of Einstein's work of 1917, *Sulla*

---

[34] D. Manetti, *Le conferenze di Alberto Einstein sulla relatività (nostra intervista col prof. Tullio Levi–Civita)*, "Il Messaggero", 28 October 1921.

[35] Einstein's nomination as a foreign member of the Academy of Sciences of Bologna was proposed by Enriques, Luigi Donati and Salvatore Pincherle. Minutes of the 2nd Extraordinary Meeting, Class of Physical Sciences, April 17, 1921, Archive of the Academy of Sciences and Institute of Bologna. In 1925 Einstein also became a foreign member (Matteucci Medal 1921) of the Accademia Nazionale delle Scienze known as Accademia dei XL in whose archives the collection "Sources for the History of Quantum Physics" is preserved.

[36] *La seconda conferenza di Einstein*, see footnote 33.

teoria speciale e generale della relatività (volgarizzazione), with a preface by Levi–Civita who wrote: "the mathematical scheme of natural philosophy has recently undergone a profound transformation by Einstein. This transformation, which he named and is now universally known as the theory of relativity, started in 1905 from the union of an experimental fact with a criterion of relativity […], [the] recognition of which rightly marks the beginning of a new era in the history of science. The successive development of Einstein's doctrines (so called general relativity) reached its maturity in 1915 thanks to the systematic employment of the methods of absolute differential calculus of our Ricci. […] The speculative importance of relativity is so enormous, that in a few years more than 700 works, books, pamphlets and articles, have been dedicated to it".[37]

Shortly thereafter, two volumes by Roberto Marcolongo were published: the first, *Relatività* (1921), collected the lectures given at the University of Naples, where he was a professor; the second, *Uno sguardo sintetico alla teoria speciale e generale della relatività* (1922), which set out to expound special and general relativity in a concise and accessible way.

A year later, again published by Zanichelli, Castelnuovo's extensive treatise on the foundations of relativity was published with the title, *Spazio e tempo secondo le vedute di Albert Einstein.*

The lectures by Luigi Donati, professor of mathematical physics, held between December 1921 and February 1922 [55–57], followed a year later by those of Paul Langevin on the structure of the atom and magnetic properties, at the Archiginnasio and the Institute of Physics, marked a continuation of the Bolognese interest in relativity. Langevin, a colleague and friend of both Enriques and Einstein, was invited as part of the initiatives of the same committee that had promoted Einstein's visit.

At that date Enriques was already working at the University of Rome. At the opening of the first of the two lectures by the French physicist, he pointed out that "the two worlds of the infinitely large and the infinitely small have always aroused the interest of mankind since antiquity" [58], and said goodbye to the Bolognese university that had hosted him as a professor for a period of thirty years, during which he had won a growing reputation in mathematics and had promoted his bold cultural activities.

A few months before the Langevin lectures, and precisely on February 8, 1923, Enriques had taken up his pen again to write to Einstein. This time, however, for serious reasons. Enriques, recalling the "unforgettable days" in Bologna and informing him of his move to Rome, reminded him of the "desire that many in Italy would have in having you permanently among us […]" also because there were rumors that conditions in [Berlin] had changed and that for reasons of anti-Semitism Einstein no longer felt at ease and was about to leave that place and Germany. "If this is so," he went on, "there the hope is reborn that we may gain you in some way for our country. […] I limited myself to talking about it with the Minister of Education, who is the idealist philosopher Prof. Gentile, and he authorized me—albeit in strict

---

[37] See the Preface by T. Levi–Civita in [3], pp. V–VII.

confidence—to tell you that he is for his part inclined to accept very willingly an initiative in this regard".[38]

Einstein, deeply moved, replied a few months later declining the invitation: "dear colleague, I was very moved by your letter, and I must confess openly that I would prefer your company, and that of Levi–Civita, to that of my colleagues here. But [...] I am very much bound to my present environment by family, friendship and business relations. At my age it is not so easy to change environment [...] But if in the future I feel forced to leave this nest of mine because of the worsening of the situation, I will immediately turn to you with joy and confidence".[39]

Great—in Adriana's recollections—was the disappointment in the building in Via Sardegna in Rome where both the Enriques and the Levi–Civitas lived; and yet it was necessary to inform the powerful minister Gentile.[40]

According to Adriana Enriques, the attempt to call Einstein to Italy was repeated ten years later. But this time—as she pointed out—it was Mussolini himself who opposed it. At that time the political atmosphere had changed profoundly since the fascist regime in 1925 had turned into a fully fledged dictatorship [28].

It was against this dictatorship, and against the appeal of Fascist intellectuals launched by Gentile at the end of a conference held in Bologna in the same year, that Einstein's Italian colleagues, in the name of freedom of thought, created the Manifesto Croce—so called because of its first signature—which was signed by illustrious representatives of the cultural and above all university world. Among the numerous adherents it is worth mentioning Guido Castelnuovo, Tullio Levi–Civita, Francesco Ruffini, Vito Volterra, Gaetano Salvemini, Paola Lombroso, Eugenio Rignano.

Despite the worsening political situation, Enriques did not quit having at least cultural contacts with Einstein, among which, first of all, was a connection with the Treccani Encyclopedia that, in that fateful year, had debuted under Gentile's direction.

In 1926 the two met in Paris; even on that occasion Enriques did not refrain from asking him for further collaborations which Einstein did not follow up. Five years later, Enriques, entrusting his thoughts to the words of the secretary of "Scientia", Paolo Bonetti, regretted not having been able to meet him in Berlin the previous December.

In 1931 the requirement to swear allegiance to the fascist regime burst upon Italian university professors with an oath that was intended to reduce the entire academic body to obedience.

Einstein's intervention against the oath of allegiance to the fascist regime was quickly publicized by his personal decision to publish, a few years later, his letter of denunciation and the plea that he sent on that occasion to the Minister of Justice Rocco, his sworn enemy since the days of the International Committee for Intellectual Cooperation.

---

[38] F. Enriques to A. Einstein, Rome 8 February 1923, Einstein Archives.

[39] A. Einstein to F. Enriques, Berlin 11 April 1923, Einstein Archives.

[40] On April 15, Enriques hastened to inform Gentile of Einstein's negative answer. See F. Enriques to G. Gentile, Rome 15 April 1923, in [59].

News of such an oath had already been circulating since 1930 by virtue of a missive from "an anonymous professor" addressed to the "Manchester Guardian" which was followed a year later (the matter had come to a temporary halt) by a letter of protest from a group of Italian intellectuals.

At the same time some important scholars in England, led by Murray who personally wrote to Rocco, in France and in Switzerland where many anti-fascists had taken refuge, had moved to support this group [60–62].

The initiative to appeal to Einstein was taken by Francesco Ruffini by virtue of their long-standing association. In the letter that reached Einstein via Geneva, thanks to the interest of the Italian exiles Guglielmo Ferrero and Mario Carrara at the head of an international movement against the oath, he informed him that neither he nor his son Edoardo intended to take the oath. Others, like for example the mathematician Volterra, would have done the same, but most would not have had the strength to refuse.

"We are left with but one hope, namely, that, if ever a voice of solidarity and protest should be raised by the most distinguished professors of foreign universities, the government will desist from its ill-considered decision, or at least will not rage against those who should refuse to take such an oath." "Judge for yourself," concluded Ruffini, "whether it is possible for you to undertake anything, to come to the aid of your colleagues in Italy".[41]

Einstein certainly didn't hold back and immediately turned to Rocco.

"My dear Sir, Two of the most eminent and respected men of science in Italy have applied to me in their difficulties of conscience and requested me to write to you with the object of preventing, if possible, a piece of cruel persecution with which men of learning are threatened in Italy. I refer to a form of oath in which fidelity to the Fascist system is to be promised. The burden of my request is that you should please advise Signor Mussolini to spare the flower of Italy's intellect this humiliation. However much our political convictions may differ, I know that we agree on one point: in the progressive achievements of the European mind both of us see and love our highest good. Those achievements are based on the freedom of thought and of teaching, on the principle that the desire for truth must take precedence of all other desires. It was this basis alone that enabled our civilization to take its rise in Greece and to celebrate its rebirth in Italy at the Renaissance. This supreme good has been paid for by the martyr's blood of pure and great men, for whose sake Italy is still loved and reverenced to-day. Far be it from me to argue with you about what inroads on human liberty may be justified by reasons of State. But the pursuit of scientific truth, detached from the practical interests of everyday life, ought to be treated as sacred by every Government, and it is in the highest interests of all that honest servants of truth should be left in peace. This is also undoubtedly in the interests of the Italian State and its prestige in the eyes of the world" so ended Einstein in his heartfelt appeal.[42]

---

[41] F. Ruffini to A. Einstein, Turin 8 November 1931, Einstein Archives, published in [60], p. 270. An excerpt from the letter is published also in [62].

[42] A. Einstein to Minister Rocco, Berlin 16 november 1931, Einstein Archives, published in [60] p. 274. See also [63].

Rocco was careful not to respond directly to Einstein, delegating this task to one of his collaborators, Giuseppe Righetti, who, in hypocritical German, admitted impassively that yes, indeed, university professors had been asked to swear an oath of loyalty to the new regime.

But—he hastened to reassure Einstein—with no demand that the professors should adhere to this or that political orientation, so much so that out of about one thousand two hundred full professors, only seven or eight had raised objections. All the others had sworn an oath; among them—Righetti maliciously specified—the "famous mathematician Levi–Civita".[43]

Einstein thus found his hands tied. His letter, unfortunately, had had no effect, and alas, Europe was heading for dark times, he commented ([60], p. 279; [32], p. 156).

In December 1931, in the French newspaper "La Liberté", but also in other newspapers published mainly in Switzerland, an article appeared denouncing the infamous oath entitled *Perché il mondo civile sappia...* written, it seems, by the exile in Paris Gaetano Salvemini, former MP, former professor of history at the University of Florence and future professor at Harvard.

Salvemini, together with Einstein and Croce, was part of a book project, elaborated in 1940 but published in the United States four years later, with the significant title, *Freedom* [64]. Salvemini contributed a careful analysis of democracy [65], while Croce devoted a long essay to the roots of freedom, developed from a historical point of view [66].

Einstein tackled a theme that was particularly dear to him, the theme of freedom and science, which he dealt with not only in relation to freedom in research, but also in relation to the freedom of science in states and society, denouncing the dangers of totalitarian regimes [67].

Free thought as an expression of reason and the material and spiritual progress of human beings was also the subject of a correspondence between Benedetto Croce and Einstein himself, deservedly published by Laterza in 1944, probably at the urging of Croce himself [68].

The first person to take up the pen at this juncture was Einstein on 7 June of that year, who expressed his most passionate solidarity with the philosopher who, at that time, was participating strenuously in the post-war political battle, rendering an "extremely precious" service to the reconstruction of Italy.[44]

In the letter, Einstein did not give up trusting in reason and therefore, following Plato, in a government ruled by philosophers. Croce responded to his "dearest letter" by first of all recalling the long fraternal conversation they had in Berlin in 1931 during which their "commonality" of democratic intentions emerged, intentions reiterated in *Freedom*. At the same time, however, the philosopher expressed to the scientist all his political reservations about the ability of philosophers to exercise rational government over world events. Rather than following Plato, it was necessary to keep in mind the teaching of Socrates, who did not hesitate to enter the political arena in times of need.

---

[43] G. Righetti to A. Einstein, Rome 12 December 1931, Einstein Archives.

[44] A. Einstein to B. Croce, Princeton 7 June 1944, in [68], pp. 3–4.

And he concluded by thanking him for his generous wishes for Italy, wishes that Einstein had made to him in the name of "the hope that his beautiful country would soon be freed from evil oppressors outside and inside".[45]

Einstein and Enriques

"The congregation of physicists shows a rather indifferent attitude towards my work on gravitation. Abraham is still the one who has shown the greatest understanding. Actually in the journal "Scientia" he curses violently against all that is relativity, but not without ingenuity".[46]

With these words Albert Einstein, writing to his friend Besso at the end of 1913, commented on the lively controversy with the German physicist Max Abraham, at that time professor of rational mechanics at the Polytechnic of Milan and an ardent anti-relativist, who between 1912 and 1914 attacked Einstein's work *in fieri* on the theory of gravitation.

From 1914 onwards Einstein began to "speak" an increasingly rich and apt Italian, since the already difficult penetration of his theory in Italy underwent a turning point and an acceleration in virtue of the controversy he engaged in with Abraham on the pages of "Scientia", the international periodical founded in 1907 under the name of "Rivista di scienza" by Federigo Enriques and Eugenio Rignano ([11], pp. 11–104).

By the spring of 1906 news of preparations had begun to circulate: "Rignano and Enriques are about to found a philosophical journal!!!".[47]

"We are all aware—Enriques wrote to his brother-in-law, the mathematician Guido Castelnuovo—of the great difficulty of the matter. We have discussed a series of coordinated themes and we propose to resort also to collaboration abroad".[48]

"Matter, energy, life, death etc.", these were the "groups of themes"[49] to be discussed in the first issues of the periodical. From the variety of these themes emerged the eclectic epistemological direction imprinted on the journal, that is to say, to avoid the dangers arising from an excess of specialisms in the name of the unity of science, and to rejuvenate the Italian cultural scene, encouraging the spread of avant-garde theories.

It is in this sense that one should read the effort of the co-editors Enriques and Rignano to establish the project of the "Rivista di scienza" ("Scientia" from 1910) with the national and international scientific community.[50]

---

[45] A. Einstein to B. Croce, Princeton 7 June 1944, cit.

[46] A. Einstein to M. Besso, (Zurich, end of 1913), in [12], p. 30.

[47] G. Vailati to G.. Papini, Basel 26 April 1906, in [69] p. 435.

[48] Cfr. F. Enriques to G.. Castelnuovo, 31 May 1906, in [70].

[49] F. Enriques to G. Vailati, 8 July 1906, in [69], p. 586.

[50] In the first issues articles appear by Émile Picard, Henri Poincaré, Charles Fabry, Svante Arrhenius, Frederick Soddy, Abel Rey, Friedrich W. Ostwald, just to name a few. In terms of the circulation of the most innovative ideas, the critical analyses of Max Planck or the reviews of William

In the first decade of the journal's life it was Enriques, above all, who dictated its agenda, choosing the themes to be investigated and discussed. From the choice of the two opening articles of the first volume, *La mécanique classique et ses approximations successives* by Picard and *Zur modernen Energetik* by Ostwald, one recognizes Enriques' unmistakable stamp, and he later succeeded in launching a debate that investigated the most recent developments in the field of classical mechanics, making use of equally illustrious collaborators such as Lorentz, Poincaré, Langevin, Abraham and Einstein.

Enriques had begun to examine the crucial questions raised in the last years of the nineteenth century by non-Euclidean geometries and regarding, above all, their foundations, when he had begun to teach at the University of Bologna, where he had arrived in 1894 appointed to teach the course in projective geometry.[51]

That year his first study of invariant geometric properties with respect to groups of transformations in the sense specified by Felix Klein's "Erlangen program" came out.

The essential guiding idea certainly did not correspond to the logical analysis of the postulates developed by Giuseppe Peano, with whom Enriques crossed swords on more than one occasion, but rather referred to "the path indicated by experimental intuition" [71], as explained by Enriques himself in an interesting correspondence with Gino Fano[52].

Contrary to Peano, but also to David Hilbert and Bertrand Russell, Enriques disliked seeing geometry dissolve into formal abstraction. On the contrary, he strove to keep geometry closely linked to the empirical reality that surrounds us and, in this perspective, he spoke several times of "physics as an extension of geometry" [73].

At the time of his arrival at the University of Bologna, he also took an interest in the study of the psychological origin of geometric postulates, which he addressed in an essay in 1901 [74].[53] From this study Enriques would take the first steps to arrive, later, at a general theory of scientific knowledge that he would call "positive gnosiology", or critical positivism.

It was thanks to the volume *Problemi della Scienza* in 1906 [76], immediately partially translated into French in 1909 and 1913 [77] and then into German and English, respectively, in 1910 [78] and 1914 [79],[54] that Enriques gained a solid international reputation not only in the field of projective geometry, but also in the field of epistemological reflection on the foundations of the physical–mathematical

---

Thomson, Robert Michels, Marie Curie and above all that of Roberto Assagioli to Sigmund Freud, which preceded the publication of two essays by Freud himself in 1913, played an important role.

[51] Enriques' work is characterized by a particular mathematical-philosophical interest in the concept of space and therefore in the epistemological questions that have developed in the field of the new non-Euclidean geometries.

[52] For the correspondence between Enriques and Fano see [72], in particular Chapt. VII "Dal calcolo geometrico a logica matematica" and Chapt. VIII "I principi della geometria e la filosofia "scientifica"", pp. 219–277.

[53] For the role of psychology in Enriques' thinking, see [75].

[54] In 1911 also a Russian translation was published by Kosmos, Maskwa, while a Spanish translation was published only posthumously in 1947 in two volumes [80].

sciences; a reflection which even Einstein had shown great "interest" in, as can be deduced from a postcard signed by himself and Heinrich Zangger before April 1920 in which, as we have said, the two asked Enriques for news about "new editions"[55] of the *Problemi della Scienza.*

Enriques' program of "considering Mechanics as an extension of Geometry" and, consequently, of proceeding to a critical analysis of the fundamental postulates of time, space, force and mass had attracted Levi–Civita's attention from the very beginning.

The central idea of Enriques' reflection—Levi–Civita explained in his review of the *Problemi* in 1907—revolved around the refusal to assume "as the basis of a definitive arrangement" of mechanics "the law of generalized inertia" [81].

It—he summarized expounding the author's point of view – "consists in extending the principle of incipient motion, with respect to a special reference system, that of fixed stars, admitting that the proportionality between force and acceleration is valid, not only at the beginning, but also throughout the course of the motion" [81].

With regard to the Enriquian criticism of the traditional concepts of Newtonian statics and dynamics, Levi–Civita stressed how *"the principle of incipient motion"* and the *"principle of composition"* do not contain "any special hypothesis about the reference system", but on the contrary "allow us to construct a (relative) statics of the material point" ([81], p. 398).

The point is – Levi–Civita went on to say—that according to Enriques "the intervention of a privileged system (that of the fixed stars), even if it is not the metaphysical absolute" seemed to him "a serious inconvenience, just as unphilosophical as a strictly geocentric conception of the universe could be" ([81], p. 400–401). This is why Enriques ended up proposing a correction of classical mechanics into a more general one, valid with respect to any reference system.

According to Levi–Civita, he was aware of the difficulties, which consisted above all in formulating "a relative dynamics" that would not get lost in "vagueness", but would allow "in the field already subjected to experiment, the same (or considerably the same) concrete predictions of Newtonian mechanics" ([81], p. 401); difficulties with respect to which Enriques showed himself to be "optimistic", encouraged by the "very new speculations on the mechanics of electrons" ([81], p. 401).

As Levi–Civita had opportunely pointed out, Enriques in 1906 discussed the principle of relativity enunciated, independently of each other, by Lorentz and Poincaré in virtue of which an electromagnetic conception of nature was delineated, whose constituent elements were no longer characterized by material particles with inertial and passive mass, but by active and dynamic electromagnetic force fields.

From this point of view, Enriques showed a special interest in the work of Lorentz who, starting from the negative result of the experiment by Albert A. Michelson and Edward W. Morley concerning the possibility of demonstrating experimentally the absolute motion of the Earth through the aether, had formulated an electric dynamics in which the mass of an electron is purely electromagnetic.

---

[55] F. Enriques to H. Zangger, Bologna 20 April 1920, Einstein Archives.

In this regard Enriques, first showed "how the explanation of the principle of relativity in the case of electro-magnetic phenomena has been obtained through the more recent developments of the theory of electrons, from which has sprung a science of electro-dynamics" ([79], p. 354). Then, after long arguments, he concluded: "Electrical dynamics is a particular case of non-Newtonian dynamics, which arises from replacing the generalized principle of inertia by a hypothesis of the solidarity of the field of motion, allowing the principles of equilibrium and of incipient motion to remain fixed ([79], p. 362).

It was precisely their expertise in this field of study that made Lorentz, Poincaré, Langevin and Brillouin the most suitable candidates to be approached to deal with these frontier themes in "Scientia" and Enriques secured their collaboration [82–85] between 1911 and 1914.

And in particular Poincaré and Lorentz, leading exponents of the electromagnetic theory of nature who, as a response to the negative outcome of Michelson and Morley's experiments, had proceeded decisively in the critical reconsideration of Newtonian mechanics, outlining, albeit from different standpoints, the principles and characteristics of a new relativistic mechanics.

As early as 1895, Lorentz realized that it was necessary to foresee a change in the temporal coordinate of an event, as well as in the spatial coordinate, when passing from one reference system to another. To this end, he introduced the notion of "local time".

In the meantime, Poincaré himself had made penetrating observations on what, in 1904, he called the principle of relativity, that is, the invariance of the laws of nature in the passage from one inertial system to another.

He, however, in an article entitled *Sur la dynamique de l'électron* [86] in 1906, published in the "Rendiconti del Circolo Matematico di Palermo", continued to reason as if the aether existed; to hold that the speed of light in vacuum was not a constant; not giving a physical interpretation of Lorentz's "local time", although he described it as an ingenious idea.

Yet, since 1902, the year of publication of his famous book, *La Science et l'hypothèse*, he had denied both the existence of absolute space and time and that of a direct intuition of the simultaneity of two events taking place in two distant places.

It was Albert Einstein who took the step that, according to an expression commonly used today, changed "the face of physics" in his article *Elektrodynamik bewegter Körper.*

Einstein operated, as is well known, a real reversal of the existing problematic framework, elevating the principle of relativity and the principle of invariance of the speed of light – which for Lorentz and Poincaré were two points of arrival— into as many postulates or axioms of a general character from which to start the mathematical construction of his theory.

As for the relations between the three protagonists of these events, Einstein and Poincaré met personally on only one occasion, at the first Solvay Congress in 1911, at the conclusion of which Einstein, somewhat disappointed, wrote to his friend Zangger that: "Poincaré was in general entirely hostile [to the theory of relativity] and, in spite of his acuity, showed little understanding of the situation" [87].

As for Lorentz, he never fully embraced Einstein's ideas, continuing to argue, as he did again at a conference held at the Teyler Foundation in Haarlem (Holland) in 1913:

"According to Einstein, it has no meaning to speak of motion relative to the ether. He likewise denies the existence of absolute simultaneity. […] As far as this lecturer is concerned, he finds a certain satisfaction in the older interpretations, according to which the ether possesses at least some substantiality, space and time can be sharply separated, and simultaneity without further specification can be spoken of" ([4], p. 166).

The difference in views did not prevent him, however, appreciating Einstein's work and establishing with him a relationship of deep mutual respect.

For his part, Enriques in an issue of "Scientia" in 1907 had published an essay entitled *Le principe d'inertie et les dynamiques non-newtoniennes* where he had summarized his ideas about the principle of inertia and non-Newtonian dynamics [88].

From this point of view, Enriques retraced the central points already exposed in the *Problemi* without, however, mentioning Lorentz or Abraham, nor Poincaré or Langevin, nor Einstein, whose name, unlike the others, did not even appear in the 1906 book.

Here—after having acknowledged Poincaré's merits for having put "The explanation of the principle of relativity in the case of electro-magnetic phenomena into a suggestive mathematical form, […]" ([79], p. 352)—he indicated among the desiderata "a more satisfactory notion of ether, which ought to be in relation to matter" ([79], p. 353), instead of being independent of it. In fact, "the hypothesis of the fixity of the ether" being "to the realm of those hypotheses which cannot fall even indirectly under the control of possible experiments" was for Enriques deprived "of all significance" and to be considered as "a mere artifice" ([79], p. 352).

We must not forget—he went on to specify – "that ether is only an intermediary of the relations between material bodies. For this reason these bodies are actually the datum, and it is proper to start from them in order to construct a notion of ether" ([79], p. 353). Which, in the last analysis, meant "defining ether as something which in every place and at every instant is in the closest connection with all matter near or distant" ([79], p. 353). And he concluded: "Without pausing over the difficulties of this programme, we confine ourselves to pointing out that the conception of ether which we require would a priori eliminate from the electro-magnetics of moving bodies, the fictitious questions which arise from adopting an absolute frame of reference. We should only need to know in case of every medium and every class of phenomena which are developed, what are the ethereal particles to which the phenomena themselves should be attributed. And if in the case of phenomena originating within the medium we have intervening only the particles that emanate from bodies belonging to that medium, the principle of relativity is assured a priori" ([79], p. 353).

Several years later, when Einstein had solved the enigma of the force of gravity with the theory of general relativity, Enriques, in an article in Spanish in 1920

[89], effectively described the reasons for the peculiar correspondence between his philosophical-scientific vision and the theory of relativity.

The latter was the confirmation that, in the scientific description of phenomena, there was a tendency towards synthesis in opposition to the analytical spirit that had characterized nineteenth-century mathematics and physics.

The adoption of non-Euclidean geometry in the field of general relativity gave geometry a new primacy in this direction; while, from the point of view of scientific knowledge, it presented itself as 'a posteriori' with respect to physics and, therefore, capable of guiding the critical analysis of physical theories.[56]

The "Scientia" Relativity Inquiry

In the meantime, in 1911, the pioneering phase of the debate of "Scientia" on relativistic concepts had begun with the essay *Il principio di relatività e i fenomeni ottici* by Castelnuovo, which, examining Einstein's ideas on special relativity, underlined their original character of great philosophical significance.

To attribute to Einstein the main merit for relativistic ideas was not from the Italian point of view a fact of little importance. In fact, when the relativistic theories began to circulate on the pages of "Scientia", in the period 1907–1908, in two distinct reviews of physics, *Le recenti teorie elettromagnetiche e il moto assoluto* and *Sulla massa elettromagnetica,* respectively, by Orso Mario Corbino and Tullio Levi–Civita, the name of Einstein was cited only at the margins of the dispute on the nature of the electromagnetic mass of the electron and, from such a perspective, lowered to the rank of simple completion of the hypotheses that Lorentz had enunciated to elaborate a theory of the electrons in motion.

The fact that in their research Lorentz and Einstein had arrived at the same formulas for expressing the mass of the electron ended up by misleading many specialists including Corbino and Levi–Civita who, at first, overlooked the conceptual and methodological differences between Lorentz's theory and Einstein's, which was considered for the most part as an interesting contribution, but secondary to the former.

So much so that Levi–Civita merely pointed out in a note "the interesting concise considerations [...] established by Mr. Einstein" [92].

As for Corbino – who in his review made a wider reference to Einstein's electromagnetic research—in dealing with the electromagnetic mass of the electron and, in particular, with the increase of the mass with the increase of its velocity, in the end sided in favor of Abraham's theory of the "rigid electron" against that of Lorentz's

---

[56] Enriques' interest in relativity and related conceptual issues is a recurrent feature of his thinking even after the *Problemi della Scienza*. His *Conferenze sulla geometria non-euclidea* held at the University of Bologna date back to 1916–17 [90], and to 1925 *Spazio e tempo davanti alla critica moderna (Questioni riguardanti le matematiche elementari)* [91].

"deformable electron"[57] since, in his opinion, this latter did not allow one "to found on pure electromagnetic bases the mechanics of the electron and therefore the whole of mechanics" [93]. "This—he continued—is instead possible with Abraham's theory, and this possibility constitutes a merit of the greatest importance from the cognitive theoretical point of view" ([93], p. 164).

To accept Abraham's theory meant – according to Corbino – defending the hypothesis of the absolute aether that "for its simplicity remains [...] today the only conceivable basis of a concretely developed electro-magnetic theory" ([93], p. 164).

And nevertheless—he concluded – "if both [...] the principle of relativity and a law of variation of the mass not responding to Lorentz's formulas were proved, one would be induced to renounce the concept of an absolute aether"; that concept that "seems absurd to Prof. Enriques" ([93], p. 165).

A few years later, to be precise in 1910, Corbino himself began to gradually change position, when he began to examine some consequences of the equivalence between mass and energy that led directly to a revision of the fundamental laws of mechanics and, therefore, to the theory of relativity.

On the other hand, after the Italian translation (1909) by Giuseppe Gianfranceschi of the famous lecture *Raum und Zeit* held in Cologne in 1908 by a former professor of Einstein's at the Zurich Polytechnic, Hermann Minkowski [94]—who had found that space and time can merge into a new four-dimensional geometry precisely from the theory of special relativity—the situation had changed: among some Italian mathematicians theoretical clarity began to be made between the positions of Lorentz and Einstein.

Besides, the development of relativistic theories, on the one hand, expressed a classical mathematical program that resorted mostly to differential equations to describe widely studied physical phenomena such as electromagnetism; on the other hand, they introduced novelties, such as non-Euclidean geometries or Minkowskian metrics, to which the mental apparatus of the mathematician of the time was sensitive and receptive.

As far as physicists were concerned, between 1906 and 1915 they showed mostly indifference to or incomprehension of the theory of relativity, and sometimes outright hostility.

This is the case, for instance, of Augusto Righi, the most authoritative Italian experimental physicist, who dealt extensively with relativity only in 1918, when he tried to explain the negative outcome of Michelson and Morley's experiment—at that time considered as the experimental basis of relativity—attributing this outcome to an erroneous interpretation of that experiment and to the poor precision of the instrumentation used.

While the editor Enriques was busy securing articles on the relativistic concepts that had become a topical subject for the issues of "Scientia", it was Einstein himself,

---

[57] Abraham had predicted a different dependence of mass on velocity for the electron from that predicted—independently—by Lorentz and Einstein. Experimental verification was first addressed in Göttingen by Walther Kauffmann and remained uncertain for a long time until another German experimental physicist, Alfred Bucherer, derived experimental data favorable to Lorentz and Einstein.

who had begun to wonder about the possibility of inserting gravitational forces into the scheme of special relativity, who entered into a controversy with Max Abraham on these themes, a controversy that, begun in 1912 in the columns of the "Annalen der Physik", appeared in 1914 in "Scientia" [95, 96].[58]

The background is well known. Abraham, starting in 1912, had elaborated with considerable mathematical skill his own theory of gravitation which was strongly competitive with that which Einstein, in turn, was laboriously developing.

Amidst harsh criticism, attacks and well-substantiated counter-attacks by both, the 1912 controversy, though officially closed by Einstein, was in fact only postponed.

The second phase of the controversy re-exploded in "Scientia" by virtue of a review entitled *Die neue Mechanik* [97], published in the January–February issue of 1914, in which Abraham stressed the superiority of Lorentz's theory of the electro-magnetic field with respect to special relativity, both on the conceptual and predictive levels. Not only that, but he pointed out the inconsistencies of special relativity, which in his opinion were already present in the formulation of 1905. He stressed the further difficulties posed by Einstein himself who, from 1911, had admitted a value for the speed of light that was no longer constant but depended on the gravitational potential.

Finally, he examined the latest elaboration of a theory of gravitation written by Einstein and Grossmann in 1913, published under the title *Entwurf einer verallge-meinerten Relativitätstheorie und einer Teorie der Gravitation* known generally as *Entwurf* [98].

In his analysis Abraham linked the mathematical difficulties encountered by Einstein and Grossmann in finding an equation of the gravitational field that satis-fied the principle of general covariance, to the physical difficulties of the theory, difficulties that he had already illustrated in the controversy of 1912 [97].

Einstein did not dodge Abraham's new attack and responded in 1914 with the essay *Zum Relativitäts-problem* in which he defended his theories, namely that of relativity in the narrow sense and that "of relativity in the broader sense [99], while pointing out the different degree of development achieved both in terms of logical consistency and from the point of view of their respective empirical corroborations.

Abraham, annoyed by the abrupt way Einstein had dismissed the discussion, replied with a critical note, *Sur le problème de la relativité*, in which he ridiculed Einstein, imagining him in the act of voting in a hypothetical referendum against special relativity [100].

Abraham's strongly polemical tones found a large following in the Italian cultural environment[59] given his adherence to the aether tradition and his ability to elaborate

---

[58] See also F. Toscano [48], esp. pp. 234–241.

[59] Also exemplary of the credit that Abraham received in Italy is the extensive and long-lasting exchange of correspondence (from 1907 to 1921) with Levi–Civita (Fondo Levi–Civita, Biblioteca dell'Accademia Nazionale dei Lincei e Corsiniana). The two had met for the first time in Rome in 1908 on the occasion of the IV International Congress of Mathematicians, where Abraham had presented a paper on the *theory of the electron* at the invitation of Levi–Civita himself. The impression was so positive that Levi–Civita himself was instrumental in winning Abraham the chair of Rational Mechanics at the Politecnico di Milano in 1909, which he held until the outbreak of war in 1914.

a revision of Newtonian gravitation theory without upsetting the concepts of classical mechanics.

Despite Einstein's scant impact on the Italian scientific community, it was precisely the diatribe of 1914 with Abraham that attracted the attention of Levi–Civita who, as mentioned above, came into contact with Einstein, becoming an important partner of his in absolute differential calculus and, in time, a strong supporter of general relativity and an indispensable interlocutor for this field of study at the international level as well.

The outbreak of the First World War had profound effects, not always scientific, on the difficult process of the acceptance of general relativity. In the bloodiest phase of the conflict, it was in fact looked upon with distrust and, paradoxically, in spite of Einstein's vigorous previously highlighted pacifist commitment, as the product of an expansionism that reflected Germany's political and military imperialism on a cultural level.

Not even "Scientia"—which passed under the exclusive editorship of Rignano after the "political" resignation of Enriques, who was against a "patriotic" management of the journal—escaped this climate.

The extent to which the influence of the anti-relativist Rignano was preponderant in the journal's choices with respect to relativity can be seen from the diatribe he had in 1919 with Attilio Palatini [101], one of Levi–Civita's most promising pupils and the author of some truly significant contributions to Einsteinian theory.

The reasons for this diatribe essentially concerned—Palatini explained in the columns of "Scientia" in the margins of an article that retraced the historical development of the theory of relativity—the anti-intuitive character of that theory, whose complex formal apparatus, if it was already difficult for mathematicians, even appeared disconcerting to those who, like Rignano, were part of a generic empiricist tradition substantially linked to so-called common sense and to experimental physics [102]. From this point of view, the theory of relativity presented itself as a collection of a priori mathematical arguments with no relationship to the physical world.

In the meantime Levi–Civita—who had moved from Padua to the University of Rome at the beginning of 1919—was trying to attract his skeptical colleagues into the relativist camp. He organized, together with Castelnuovo and the mathematician Vito Volterra, a series of lectures on general relativity that were held between March and April at the Mathematical Seminar in Rome of which Volterra was director [103].

The strategy adopted by Levi–Civita was to entrust an introductory historical lecture on absolute differential calculus to the expository clarity of the Neapolitan mathematician Roberto Marcolongo [104].[60]

Equally astute was the choice adopted by Levi–Civita for the title of his lecture, *Come potrebbe un conservatore giungere alla soglia della nuova meccanica*, in which he proposed to give prominence to those elements of the Newtonian mechanics that would allow a gradual approach to the Einsteinian theory of gravitation.

---

[60] On 10 April 1919, Marcolongo held a further lecture titled *La teoria della relatività in senso stretto* ([104], pp. 55–76).

It was the success of the astronomical expeditions of 1919 that provided the decisive breakthrough for the experimental confirmation of Einstein's theory. Until then, attempts to measure the supposed deflection of light in a gravitational field during a solar eclipse had come to nothing. In March 1917, it was the astronomer royal Frank Watson Dyson who drew attention to the favorable conditions that would occur at the eclipse of May 29, 1919, and organized with Sir Arthur Eddington, director of the Greenwich Observatory, two separate expeditions, one to Principe Island in the Gulf of Guinea directed by the latter, the other to Sobral, Brazil. The results were so convincing that they were presented in London to the members of the Royal Society and the Royal Astronomical Society on November 6, 1919, during the famous meeting mentioned in the first paragraph.

In the same year—within the International Astronomical Union of the Conseil international des recherches, a transnational research body established in Brussels by the victorious countries of the First World War and of which Volterra was one of the vice-presidents—a commission for the study of relativity was established. Chaired by Eddington, its members included the Italian Levi–Civita.[61]

The first meeting was held in Italy, in Rome, in May 1922, on the occasion of the general assemblies of the International Geodetic-Geophysical and Astronomical Unions, organized personally by Volterra with the scientific and logistical support of the Accademia Nazionale dei Lincei, of which he had meanwhile become vice-president.

In the absence of Eddington, it was Volterra himself who delivered the inaugural speech in which, if, on the one hand, he underlined how "from the mathematical, i.e. logical, point of view, the theory of relativity is perfect, just as the ordinary Newtonian theory is perfect, just as any other theory resting on non-contradictory postulates would be mathematically perfect"; on the other hand, he noted that "if the theory agrees with some fact that rebels against the Newtonian theory and the new astronomical facts are verified, this theory will constitute a mathematical armor that will adapt itself better to the phenomena of astronomy [...] and therefore it will be preferred to the other".[62]

The measured words of a valuable scientist such as Volterra were countered by the obtuse objections of the president of the Italian Astronomical Society, Vincenzo Cerulli, who ventured a hasty judgement of relativity seen as "a degenerative crisis" [106] of physics, anticipating arguments that, shortly afterwards, would hold sway again in "Scientia".

---

[61] The Commission in 1919 also included M. Brillouin, H.D. Curtis, Th. De Donder, J. Ishiwara, J. Jeans, E. Picard. In 1922 Langevin also joined. In 1925, the year in which the Executive Committee of the International Astronomical Union decided to abolish the Commission, the members were: Levi–Civita, President; Angelitti, Armellini, Birkhoff, E. Borel, De Sitter, De Donder, Eddington, Picard, J.M. Plans, L. Silberstein, R.C. Tolman. For an initial account of these events we refer to [105].

[62] V. Volterra, "Discorso congresso Astronomico, Geodetico, Geofisico (1922)", Vito Volterra Archive, Serie III—Attività scientifica, busta 105, Biblioteca dell'Accademia Nazionale dei Lincei e Corsiniana, Rome.

The need to discuss the nature of Einstein's theory without indulging in sterile polemics, but rather by resorting to epistemological analysis, was the basis of a second "great enquiry of clarification, criticism and evaluation" ([107], p. 13) planned by the editor Rignano in the autumn of 1922.

The investigation began in 1923 with an essay entitled *La question préalable contre la théorie d'Einstein* [107] by the Frenchman Henri Bouasse [108, 109], an expert in acoustics and optics, but above all, in this case, a tenacious anti-relativist. As an expert, however, he was disappointing since his essay exhibited an absolute lack of content from the very beginning: all the reasoning and results of relativity were wrong, he asserted bluntly.

The second article, *Lo spazio-tempo dei relativisti ha un contenuto reale?* (1923), written by Guido Castelnuovo who supported a "realist" conception of space–time maintaining that between a four-dimensional geometric representation of space–time and real space–time there is the same relationship that exists between a geographical map and the territory it represents, was of an entirely different tenor.

As evidence of the international capacity for aggregation of "Scientia", many foreigners were recruited by Rignano for the survey. In fact, in 1923 the paper *Can Gravitation be explained?* appeared, written by the famous astronomer protagonist of the 1919 expedition, Arthur Eddington, a long-time collaborator of "Scientia" [110],[63] who had tried his own specific and original interpretation of relativity, seen as a free creation of the human mind based on a priori principles that can be applied to the real world through mental experiments and a small number of elementary empirical data.

In 1924, Hans Reichenbach, the future founder in 1928 of the Berlin Society for the Philosophy of Science and a distinguished representative of neo-positivism, intervened in the debate by analyzing with solid arguments the philosophical significance of the foundations of the theory of relativity.

From this point of view, an extremely relevant point highlighted in his article in "Scientia", *Die relativistische Zeitlehre*, concerned the presence, in Einstein's definition of time, of a factor that was arbitrary but positive, since it exercised a real descriptive function. In this regard, he pointed out that our definition of time concerns only the *interpretation* of our perceptions and not the perceptions themselves, and that, therefore, arbitrariness in the choice of simultaneity *never* leads to a false statement about what has actually been observed [112].

In the course of the survey, among the Italians, the debate on the foundations continued to be discussed especially by the pro-relativists Guido Fubini, Roberto Marcolongo and Francesco Severi [113–116].

In 1923, at the same time as the Scientia inquiry, a collection of essays by Italian scholars was published as an appendix to the Italian translation of the popularizing treatise by the German astronomer August Kopff, *I fondamenti della relatività einsteiniana*, of 1923 [117].

---

[63] The association between Eddington and "Scientia", begun in 1910 with his article *Star-Streams*, had continued with *The stellar Universe as a Dynamical System* (1915) and *The Interior of a Star* (1918). On the reception of relativity in English circles see [111].

In the context of the review, it was mainly astronomers who took up positions that were strongly skeptical or prejudicially hostile to the theory, opting for an all-out defense of the classical law of gravitation.[64] On the other hand, it was Enrico Fermi, at the age of twenty, who expressed an original thought, grasping the most significant developments of the theory of relativity from a physical point of view. Uninterested in the epistemological value of science, Fermi was particularly interested in the promising theoretical consequences of relativity for the study of the structure of the atom by virtue of the equivalence between mass and energy established by Einstein. Experimentally, the famous formula $E = mc^2$ predicted exceptional results for the "frightening amount of energy" that could be released "in the near future" [119].

The controversy on relativity ended rather quickly and with it the raison d'être of the investigation of "Scientia" which closed in 1926.

The attention of scientists in fact began to be directed towards new directions of research: the theories of the physics of the nucleus; the study of the evolution and of the structure of the universe; the development of unitary theories of the electromagnetic and gravitational field.

The feeling that one obtains is that interest in relativity began to wane as the theory was consolidated, reaching a degree of reliability equal to that of the other branches of mathematical physics, as witnessed, for example, by Orso Mario Corbino in 1927 in the pages of "Il Nuovo Cimento" [120], in which he identified the new ground for comparison in quantum mechanics and, in particular, in the opposite interpretations of it given by Heisenberg-Born-Jordan with matrix mechanics and Erwin Schrödinger with his wave mechanics.

While the enquiry of "Scientia" was in its final stages and questions lingered on that were by now obsolete, Einstein had gone much further in his search for a unity of nature, trying, from the mid 1920s, to elaborate a field theory that unified gravitation and electromagnetism.

Only in 1931 was a long popularizing article published in "Scientia" on the major scientific syntheses elaborated over the past few years by Einstein, Werner Heisenberg, Schrödinger and Paul Dirac, edited by the mathematical physicist of the University of Genoa, Paolo Straneo, who had graduated in 1896 in engineering at the Swiss Federal Polytechnic of Zurich during the same period as Besso. He had been in correspondence with Einstein since 1915 on the subject of gravitation, then, for years, he worked hard on unitary theories, elaborating his own version of them, at first sight sufficiently convincing in the choice of the mathematical formalism to be used, but in fact incapable of achieving the objectives he had set himself.

As is well known, the probabilistic interpretation of quantum mechanics – proposed by Heisenberg and Bohr—according to which the laws of the microscopic world are not deterministic, had triggered heated debates within the scientific community, not only of a technical nature but also of a philosophical order. Einstein himself had no confidence in the probabilistic interpretation because of its statistical character and, on this point, had engaged in a heated discussion with Niels Bohr on the

---

[64] On the other hand, it was the astronomers of Stockholm who contested the attribution of the Nobel Prize for relativity to Einstein and thus blocked the process of the prize in 1921. See [118].

foundations of physics. "Despite its great successes, I am rather skeptical towards quantum mechanics", Einstein wrote in 1928 to Straneo, strongly denouncing how "the renunciation of causality is unbearable for me".[65]

In agreement with Einstein and in favor of a causalistic and deterministic vision of physical reality was Enriques who, during the Florentine Congress of the Society 'Mathesis' in 1930, invited Enrico Persico and Enrico Fermi—supporters of a resolutely antideterministic point of view—to expound quantum mechanics and to discuss its interpretation. "There exists a grand new edifice of theoretical physics which finds its highest expression today in quantum mechanics. Prof. Fermi has invited us to give up the classical concept of determinism, paying this price for the success of the new doctrine. Some would perhaps be willing to consent to this; but they will no longer pay if they see that one can enter the theater without paying: the fates of quantum mechanics and indeterminism are not necessarily in agreement" [121].

That same year, after Rignano's death, Enriques returned to edit "Scientia", beginning a new phase for the journal aimed at discussing the most burning scientific problems. In April 1930, he wrote to his famous friend[66] Albert Einstein, not only soliciting his collaboration with the journal, but also suggesting the controversial subject of the "indeterminism of Quantum Mechanics"[67] among the possible topics on which to intervene! Unfortunately, Einstein did not respond in the way Enriques had hoped.

During the Thirties, the various issues of "Scientia", were cleverly adapted by the editor preponderantly towards the physics of the nucleus in order to deepen the probabilistic-indeterministic interpretation of quantum mechanics.

It is not surprising, therefore, to find, among the scientists recruited by Enriques for the debate on the foundations of physics and science in general, some of the most prestigious scientist-philosophers, mostly linked to the Vienna Circle, such as Moritz Schlick, who was one of the founders of the Circle, Philipp Frank, Rudolf Carnap and Otto Neurath.[68]

Some time earlier, Neurath himself, the main inspirer of the manifesto of the Circle published in 1929 with the solemn title *Wissenschaftliche Welauffassung* [122], had counted Enriques and Einstein among the illustrious ancestors of the association in

---

[65] A. Einstein to P. Straneo, 8 May 1928, Einstein Archives.

[66] During the 1930's Enriques continued to deal with relativity within the history of scientific thought as, for example, in the introduction to the 1930 reprint of P. Tannery's *Pour l'Histoire de la science hellène*, in *La teoria della conoscenza scientifica da Kant ai nostri giorni* (1938) as well as in the *Compendio di storia del pensiero scientifico*, written with G. de Santillana in 1937, where the much discussed experimental confirmation of Einstein's theory provided by the shift of Mercury's perihelion is indicated as an established fact.

[67] F. Enriques to A. Einstein, 15 April 1930, The Albert Einstein Archives, The Jewish National & University Library, The Hebrew University of Jerusalem.

[68] M. Schlick, *Erkenntnistheorie un moderne Physik* (1929); Ph. Frank, *Der Charakter der heutigen physikalischen Theorie* (1931), *Positivistische oder metaphysische Auffassung der Physik?* (1935); R. Carnap, *Existe-t-il des prémisses de la science qui soient incontrôlables?* (1936); O. Neurath, *Physikalismus* (1931), *La notion de "type" à la lumière de la logique nouvelle* (1937), *Die neue Enzyklopädie des wissenschafthchen Empirismus* (1937).

relation to the foundations, aims and methods of empirical science, together with Helmholtz, Riemann, Mach, Duhem, Boltzmann and Poincaré.

In 1937, Neurath, favorably impressed by the speech given by Enriques two years earlier on the occasion of the International Congress of Scientific Philosophy in Paris [123], invited him to become a contributor to the *International Encyclopedia of Unified Science*—an initiative presented as the main instrument for the development of the movement for the unity of science. He asked him, among other things, to write some introductory pages on the unity of science for the first issue of the *Encyclopedia* as well as to compile a specific one on the history of science.

Enriques agreed to both requests, which, alas, he did not complete. Shortly after, in fact, his personal and professional projects were overwhelmed by the racial laws of the Fascist regime, as were those of several of the protagonists of the *Encyclopedia*. In 1939, Neurath informed Carnap of what had happened: "Enriques is no longer co-editor of "Scientia". This is the effect of his Jewish origins".[69]

Enriques, although expelled from the university, banished even from the library of the Faculty of Science together with Castelnuovo, ousted from all cultural societies and academies and forced to abandon the editorship of "Scientia", did not give up and managed to maintain some cultural visibility by publishing abroad or under a false name.

Across the ocean, at Princeton, Einstein continued to pursue his research and pacifist campaigns by helping exiled Jewish colleagues to rebuild a new and academic existence. Since 1930, he had already acted in this way by responding promptly and positively to the request of two old friends, Enriques and Levi–Civita, to support the academic career of Alfred Rosenblatt, a Polish Jew, a mathematician at the University of Krakow, who was seeking refuge in South America.[70]

Nor did he forget to "speak Italian" when it came to defending these and others of "his" colleagues from bullying and dictatorship in the name of "human freedom" and the "search for scientific truth".[71]

At the end of the war, having survived suffering and persecution, while Einstein rejoiced with "his" colleague, the new president Castelnuovo, for the rebirth of the Accademia dei Lincei after the Fascist oppression and renewed his desire "to become again a *foreign associate* of your Academy as I have been in the good times of the past",[72] Neurath wrote to Enriques to rekindle the fabric of a friendship and collaboration so violently broken, in the hope that the letter could reach him "as a dove of peace".[73]

---

[69] O. Neurath to R. Carnap, 11 January 1939, in [124].

[70] See F. Enriques to A. Einstein, Rome 20, 27 February and 15 April 1930, Einstein Archives.

[71] A. Einstein to Minister Rocco, Berlin 16 November 1931, Einstein Archives, also in [60] p. 274.

[72] A. Einstein to G. Castelnuovo, 26 June 1946, Archivio storico della Reale Accademia dei Lincei, pos. 4, fasc. "Soci stranieri riconfermati aprile 1946-ottobre 1947", Accademia Nazionale dei Lincei, Rome.

[73] O. Neurath to F. Enriques, 22 September 1945, Otto Neurath Korrespondenz, Rijksarchief, Haarlem (Noord-Holland Archiev).

References

1. A. Einstein, *Grundlage der allgemeinen Relativitätstheorie.* "Annalen der Physik" IV, 49, no. 7, 769–822 (1916)
2. A. Einstein, *Über die Spezielle und die Allgemeine Relativitätstheorie, Gemeinverständlich* (F. Vieweg, Branschweig, 1917)
3. A. Einstein, *Sulla teoria speciale e generale della relatività (volgarizzazione).* Preface by T. Levi–Civita, (Zanichelli, Bologna, 1921)
4. A. Pais, *Subtle is the Lord. The Science and the Life of Albert Einstein* (Oxford University Press, Oxford-NewYork-Toronto-Melbourne, 1982)
5. M.M.G. Ricci, T. Levi–Civita, *Méthodes de calcul differentiel absolu et leur applications.* "Mathematische Annalen" 54, 125–201 (1900)
6. A. Einstein, *Über einen die Erzeugung und Verwandlung des Lichtes betreffenden heuristischen Gesichtspunkt.* "Annalen der Physik", IV, 17, 132–148 (1905)
7. A. Einstein, *Die von der molekularkinetischen Theorie der Wärme gefordete Bewegung von in ruhenden Flüssigkeiten suspendierten Teilchen.* "Annalen der Physik", IV, 17, 549–560 (1905)
8. A. Einstein, *Elektrodynamik bewegter Körper.* "Annalen der Physik", IV, 17, 891–921 (1905)
9. A. Einstein, *Ist die Trägheit eines Körpers von seinem Energienhalt abhängig?.* "Annalen der Physik", IV, 18, 639–641 (1905)
10. R. Simili (ed.), *Federigo Enriques. Per la scienza. Scritti editi e inediti* (Bibliopolis, Naples, 2000)
11. S. Linguerri, *La grande festa della scienza. Eugenio Rignano and Federigo Enriques. Lettere* (Franco Angeli, Milan, 2005)
12. P. Speziali, Introduction, in *Albert Einstein Correspondence avec Michele Besso (1903–1955)* (Herman, Paris, 1979), pp. LXII
13. R. Marcolongo, *Davide Besso,* "Periodico di Matematiche", ser. 3, n. 4, 147–156 (1907)
14. P. Dossier, *Michel Besso 1873–1955.* "Archives des Sciences", IX, 73–76 (1956)
15. A. Caracciolo, *Una diaspora da Trieste: I Besso nell'Ottocento.* "Quaderni Storici", XVIII, 54, 3, 897–912 (1983)
16. H.A. Medicus, *The Friendship among Three Singular Men. Einstein and His Swiss Friends Besso and Zangger.* "Isis", 85, 456–478 (1994)
17. R. Schulmann, A.J. Kox, M. Janssen, J. Illy (eds), *The Collected Papers of Albert Einstein,* vol. 8: *The Berlin Years. Correspondence, 1914–1918, Part A: 1914–1917* (Princeton University Press, Princeton, 1998), p. 572
18. P. Frank, *Einstein: His Life and Times* (Jonathan Cape, London, 1948)
19. J. Renn, R. Schulmann (eds.), *Albert Einstein, Mileva Marić: The Love Letters, translated by Shawn Smith* (Princeton University Press, Princeton, 1992)
20. A. Einstein, *Preface,* in *Cinquant'anni di relatività 1905–1955,* ed. by M. Pantaleo (Edizioni Giuntine & Sansoni editore, Florence, 1955), p. XVIII
21. C. Bracco, *Quand Albert devient Einstein* (CNRS Éditions, Paris, 2017)
22. B. Hoffmann, H. Dukas, *Albert Einstein Creator and Rebel* (Granada Publishing, 1975), p. 26
23. M. Winteler-Einstein, *Albert Einstein—Beitrag für sein Lebensbild (Excerpt),* in *The Collected Papers of Albert Einstein,* vol. 1: *The Early years, 1879–1902,* ed. by J. Stachel et al. (Princeton University Press, Princeton, 1987), pp. XLVIII–LXVI
24. E. Pelizza Marangoni, *Momenti pavesi nella vita di Alberto Einstein,* "La Provincia Pavese", 14 May (1955), p. 1
25. E. Sanesi, *Three letters by Albert Einsten and some informations on Einstein's stay at Pavia.* "Physis", XVIII, 2, 176–177 (1976)
26. L. Fregonese (ed.), *Gioventù felice in terra pavese* (Cisalpino Editore, Milan, 2005), p. 16
27. A. Enriques, *Einstein poteva insegnare all'Università di Roma,* "L'Europeo", 1 May (1955), p. 14
28. M. Chierici, *Lessico famigliare/La figlia del matematico Federigo Enriques apre l'album dei ricordi. "Io e Einstein alla stazione..."*, "La Stampa", 10 April (1992), p. 9

29. H. Dukas, B. Hoffmann (eds.), *Albert Einsten, the Human Side: New Glimpses from his Archives* (Princeton University Press, Princeton, 1979), p. 151
30. E. Ferrero, *In un diario le grandi firme del nostro secolo,* "La Stampa", 12 August (1988), p. 17
31. B. Schroeder-Gudehus, *Les scientifiques et la paix: la communauté scientifique internationale au cours des années 20* (Presses de l'Université de Montréal, Montréal, 1978)
32. O. Nathan, M. Norden, *Einstein on Peace* (Simon and Schuster, New York, 1960)
33. A.N. Whitehead, *Science and the Modern World* (Cambridge University Press, Cambridge, 1929), p. 13
34. A.N. Whitehead, *The Principle of Relativity with Applications to Physical Science* (Cambridge University Press, Cambridge, 1922)
35. A. Pais, *Einstein Lived Here: Essays for the Layman* (Clarendon Press-Oxford University Press, Oxford, New York, 1994)
36. R.M. Friedman, *The Politics of Excellence: Behind the Nobel Prize in Science* (H. Holt & Company, New York, 2001)
37. J.R. Goodstein, *The Volterra Chronicles: The Life and Times of an Extraordinary Mathematician 1860–1940* (American Mathematical Society, Providence, 2007)
38. G. Paoloni, R. Simili, *Vito Volterra and the making of research institutions in Italy and abroad,* in *The migration of ideas,* ed. by R. Scazzieri, R. Simili (*Science history publications, Sagamore Beach,* 2008), pp. 123–150
39. A. Guerraggio, G. Paoloni, *Vito Volterra* (Springer, Berlin, Heidelberg, 2012)
40. D.K. Buchwald et al. (eds.), *The Collected Papers of Albert Einstein,* vol. 12: *The Berlin Years. Correspondence,* January-December 1921 (Princeton University Press, Princeton, 2009)
41. G. Enriques, *Via d'Azeglio 57* (Zanichelli, Bologna, 1983)
42. R. Maiocchi, *Einstein in Italia. La scienza e la filosofia italiane di fronte alla teoria della relatività* (Franco Angeli, Milan, 1985)
43. R. Simili, *Einstein a Bologna,* in *Albert Einstein, ingegnere dell'universo,* ed. by F. Bevilacqua, J. Renn (Skira Editore, Milan, 2005), pp. 284–288
44. R. Simili (ed.), *Federigo Enriques, filosofo e scienziato* (Cappelli, Bologna, 1989)
45. L. Quilici, R. Ragghianti, *Il carteggio Xavier Léon: corrispondenti italiani con un'appendice di lettere di Georges Sorel,* "Giornale critico della filosofia italiana", LXVIII (1989), pp. 295–368, esp. pp. 311–315
46. F. Enriques, *Un convegno di matematici e filosofi,* "Il Marzocco", XIX, 8 March (1914), p. 2
47. T. Levi–Civita, *Commemoration of Gregorio Ricci-Curbastro,* "Memorie Accademia dei Lincei", ser. VI, I (1925), pp. 555–564, also in G. Ricci-Curbastro, *Opere matematiche,* Vol. I (Unione Matematica Italiana, Rome, 1956–1957), pp. 2–3
48. F. Toscano, *Il genio e il gentiluomo. Einstein e il matematico italiano che salvò la teoria della relatività generale* (Sironi Editore, Milan, 2004)
49. J. Goostein, *Einstein's Italian Mathematicians: Ricci, Levi–Civita, and the Birth of General Relativity* (American Mathematical Society, Providence, 2018)
50. E. Sangiorgi, *Albert Einstein e Gregorio Ricci-Curbastro: Serendipity?,* "Quaderni di Storia della Fisica", 20, no. 1, 27–41 (2018)
51. L. Fregonese, *Il giovane Albert Einstein a Pavia,* "Giornale di Fisica", 59, no. 1, 47–71 (2018)
52. *Atti del IV Congresso internazionale di filosofia (Bologna 1911)* (Formiggini, Genoa, 1911)
53. R. Simili, *Sotto i riflettori: Enriques, Einstein e il IV Congresso internazionale di filosofia,* in *Una scienza bolognese. Figure e percorsi nella storiografia della scienza,* ed. by A. Angelini, M. Beretta, G. Olmi (Bononia University Press, Bologna, 2015), pp. 243–262
54. F. Enriques, *Le conferenze di Alberto Einstein a Bologna,* "Rivista filosofica", II, 271–274 (1921), also in R. Simili (ed.), *Federigo Enriques. For science. Scritti editi e inediti,* cit., p. 329
55. L. Donati, *Introduzione alla teoria della relatività,* "L'Elettrotecnica", IX, 7, 147–149 (1922)
56. L. Donati, *La relatività speciale,* "L'Elettrotecnica", IX, 13, 286–289 (1922)
57. L. Donati, *La relatività generale,* "L'Elettrotecnica", IX, 18, 401–402 (1922)
58. *Le conferenze di Paul Langevin all'Archiginnasio e all'Istituto di Fisica,* "L'Archiginnasio," XVIII, 101 (1923)

59. A. Guerraggio, P. Nastasi (eds.), *Gentile e i matematici italiani. Lettere 1907–1943* (Bollati Boringhieri, Turin, 1993), pp. 150–151

60. L. Polverini, *Albert Einstein e il giuramento fascista del 1931.* "Rivista storica italiana" 103, 1, 268–280 (1991)

61. A. Einstein, *The World as I See It, translated from the German by A. Harris* (John Lane, The Bodley Head, London, 1935)

62. H. Goetz, *Der freie Geist und seine Widersacher: die Eidverweigerer an den italienischen Universitaten im Jahre 1931* (Haag und Herchen, Frankfurt am Main, 1993)

63. A. Einstein, *Ideas and Opinions, based on Mein Weltbild (The world as I see it), edited by Carl Seeling and other sources, new translations and revisions by Sonja Bargmann* (Wings Books, New York, 1954), p.23

64. R.N. Anschen (ed.), *Freedom, Its Meanings* (Harcourt Brace & Co., New York, 1940)

65. G. Salvemini, *Democracy Reconsidered,* in *Freedom, Its Meanings,* cit., pp. 329–348

66. B. Croce, *The Roots of Liberty,* in *Freedom, Its Meanings,* cit., pp. 24–41

67. A. Einstein, *Freedom and Science,* in *Freedom, Its Meanings,* cit., pp. 381–383

68. A. Einstein, *Lettera a B. Croce e risposta del Croce* (Gius. Laterza & Figli, Bari, 1944)

69. G. Vailati, *Epistolario 1891–1909,* ed. by G. Lanaro (Einaudi, Turin, 1971)

70. U. Bottazzini, A. Conte, P. Gario (eds.), *Riposte armonie. Lettere di Federigo Enriques a Guido Castelnuovo* (Bollati Boringhieri, Torino, 1996) pp. 639–640

71. F. Enriques, *Memorie scelte di geometria,* vol. I (Zanichelli, Bologna, 1956), p. 161

72. U. Bottazzini, *Va' Pensiero. Immagini della matematica nell'Italia dell'Ottocento* (Il Mulino, Bologna, 1994)

73. G. Israel, *Poincaré et Enriques. Deux conceptions différentes sur les rapports entre mathématique, mécanique et géométrie,* in *1830–1930: A Century of Geometry, Epistemology, History and Mathematics,* ed. by L. Boi, D. Flament, J.M. Salanskis (Springer, Berlin, 1992), pp. 107–126

74. F. Enriques, *Sulla spiegazione psicologica dei postulati della geometria,* "Rivista filosofica", III, n. 4, 171–195 (1901); also in F. Enriques, *Natura, ragione e storia* (Einaudi, Turin, 1958), pp. 71–94, and in *Memorie scelte di geometria,* cit. II, pp. 145–161

75. G. Israel, *Federigo Enriques: a psychologistic approach for the working mathematician,* in *Perspectives on psychologism,* ed. by M.A. Notturno (Brill, Leiden, 1989), pp. 426–457

76. F. Enriques, *Problemi della Scienza* (Zanichelli, Bologna, 1906)

77. F. Enriques, *Les Problèmes de la science et de la logique,* part I (chapts. I–III) (Alcan, Paris, 1909) and *Les concepts fondamentaux de la science. Leur signification réelle et leur acquisition psychologique,* part II chapts. IV–VI) (Flammarion, Paris, 1913)

78. F. Enriques, *Probleme der Wissenschaft,* 2 vols.: *1. Wirklichkeit und Logik, 2. Die Grundbegriffe der Wissenschaft* (Teubner, Leipzig, 1910)

79. F. Enriques, *Problems of Science* (Open Court Co., London-Chicago, 1914)

80. F. Enriques, *1. Problemas de la lógica,* (chapts. I–III), *2. Problemas de la ciencia* (chapts. IV–VI) (Espasa-Calpe Argentina, Buenos Aires-Mexico, 1947)

81. T. Levi–Civita, *Le idee di Enriques sui principî della meccanica.* "Rivista di filosofia", XVI, 337–346 (1907); also in R. Simili (ed.), *Federigo Enriques. Per la scienza. Scritti editi e inediti,* cit., p. 400

82. P. Langevin, *L'évolution de l'espace et du temps.* "Scientia", 10, 31–54 (1911)

83. H. Poincaré, *L'espace et le temps.* "Scientia", 12, 159–171 (1912)

84. M. Brillouin, *Propos sceptiques au sujet du principe de relativité.* "Scientia", 13, 10–26 (1913)

85. H.A. Lorentz, *La gravitation.* "Scientia," 16, 28–59 (1916)

86. H. Poincaré, *Sur la dynamique de l'électron.* "Rendiconti del Circolo Matematico di Palermo" XXI, 129–175 (1906)

87. M.J. Klein, A.J. Kox, R. Schulmann (eds.), *The Collected Papers of Albert Einstein.* vol. 5: *The Swiss Years. Correspondence, 1902–1914,* ed. by M.J. Klein, A.J. Kox, R. Schulmann (Princeton University Press, Princeton, 1993), p. 349

88. F. Enriques, *Le principe d'inertie et les dynamiques non-newtoniennes.* "Rivista di scienza" 2, 21–34 (1907)

89. F. Enriques, *La evolución del concepto de la géométría y la escuela italiana durante los últimos cincuenta anos.* "Revista Matemática Hispano-Americana", II, 1–2, 1–17 (1920)

90. F. Enriques, *Conferenze sulla geometria non-Euclidea*, ed. by O. Fernandez (Zanichelli, Bologna, 1918)

91. F. Enriques, *Spazio e tempo davanti alla critica moderna (Questioni riguardanti le matematiche elementari)* (Zanichelli, Bologna 1924–27), pp. 429–459

92. T. Levi–Civita, *Sulla massa elettromagnetica.* "Rivista di scienza" 2, 387–412 (1907), esp. 407

93. O.M. Corbino, *Le recenti teorie elettro-magnetiche e il moto assoluto.* "Rivista di scienza" 1, 160–167 (1907), esp. 161

94. H. Minkowski, *Raum und Zeit.* "Physikalische Zeitschrift" XX, 104–111 (1909)

95. M. De Maria, *Le prime reazioni alla relatività generale in Italia: le polemiche fra Max Abraham e Albert Einstein, in La matematica italiana tra le due guerre mondiali (Atti del Convegno, Milano–Gargnano del Garda, 8–11 October 1986)* (Pitagora Editrice, Bologna, 1987), pp.143–159

96. C. Cattani, M. De Maria, *Max Abraham and the Reception of Relativity in Italy: His 1912 and 1914 Controversies with Einstein*, in *Einstein and the History of General Relativity*, ed. by D. Howard, J. Stachel (Birkhäuser, Boston, MA, 1989), pp. 160–175

97. M. Abraham, *Die neue Mechanik.* "Scientia", 15, 8–27 (1914)

98. A. Einstein, M. Grossmann, *Entwurf einer verallgemeinerten Relativitäts theorie und einer Teorie der Gravitaion. I. Physikalisher Teil von Albert Einstein. II. Mathematischer Teil von Marcel Grossmann* (G. Tubner, Leipzig und Berlin, 1913). Reprinted with the addition of *Bemerkungen*, "Zeitschrift für Mathematik und Physik", 62, 225–261 (1914)

99. A. Einstein, *Zum Relativitäts-problem.* "Scientia", 15, 337–348 (1914)

100. M. Abraham, *Sur le problème de la relativité.* "Scientia", 16, 101–103 (1914)

101. Cf. C. Cattani, *Levi–Civita's influence on Palatini's contribution to General Relativity*, in *Einstein and the History of General Relativity*, cit., pp. 206–222

102. A. Palatini, *La teoria della relatività nel suo sviluppo storico*, Part II, *La relatività generale.* "Scientia", 26, 277–289 (1919)

103. Cf. "Rendiconti del Seminario matematico della Facoltà di Scienze della R. Università di Roma 1918–1919" (Rome, 1920)

104. R. Marcolongo, *I fondamenti analitici della teoria generale della relatività e le equazioni del campo gravitazionale.* "Rendiconti del Seminario Matematico della Facoltà di Scienze dell'Università di Roma", 5, 77–94 (1918–1919)

105. G. Maltese, *The late entrance of relativity into Italian scientific community* (1906–1930), "Historical Studies in the Physical Biological Science", 31, 1, 125–173 (2000)

106. V. Cerulli, *Address by Professor Cerulli*, "Transactions of the International Astronomical Union," I, 146 (1922)

107. H. Bouasse, *La question préalable contre la théorie d'Einstein.* "Scientia", 33, 13–24 (1923)

108. Cf. M. Paty, *The Scientific Reception of Relativity in France*, in *The Comparative Reception of Relativity*, ed. by Th. F. Glick (D. Reidel Publishing Company, Dordrecht, 1987), pp. 147, 150

109. M. Biezunski, *Einstein's Reception in Paris in 1922*, in *The Comparative Reception of Relativity*, cit., p. 176

110. A.S. Eddington, *Can Gravitation be explained?.* "Scientia", 33, 313–324 (1923)

111. J.M. Sánchez-Ron, *The Reception of Special Relativity in Great Britain*, in *The Comparative Reception of Relativity*, cit., pp. 27–58

112. H. Reichenbach, *Die relativistische Zeitlehre.* "Scientia", 36, 361–374 (1924)

113. G. Fubini, *Sul valore della teoria di* Einstein. "Scientia", 35, 85–92 (1924)

114. R. Marcolongo, *La relatività ristretta. Parte Prima: Suo punto di partenza sperimentale.* "Scientia", 35, 249–258 (1924); *Parte Seconda: Modificazioni che essa apporta nei concetti di spazio e di tempo.* "Scientia", 35, 321–330 (1924)

115. F. Severi, *Elementi logici e psicologici dei principi di relatività.* "Scientia", 37, 1–10 (1925)

116. F. Severi, *Esame delle obiezioni di ordine generale contro la relatività del tempo.* "Scientia", 37, 77–86 (1925)

117. R. Contu, T. Bembo (eds.), *Augusto Kopff. The foundations of Einsteinian relativity. Value and interpretation of the theory in the original writings of A. Aliotta, E. Bianchi; G. Boccardi, A. Bonucci, P. Burgatti, V. Cerulli, P. Emanuelli, F. Enriques, E. Fermi, G. Fubini, G. Gianfranceschi, M. La Rosa, Q. Majorana, E. Rignano, U. Spirito, A. Tilgher, E. Troilo, É. Borel, H. Weyl* (Ulrico Hoepli, Milan, 1923)
118. K. Grandin, *The difficult task to award Einstein a Nobel Prize.* "Il Nuovo Saggiatore", 37, no. 1–2, 7–16 (2021)
119. E. Fermi, *Le masse nella teoria della relatività*, in *Augusto Kopff. I fondamenti della relatività einsteiniana*, cit.
120. O.M. Corbino, *La crisi odierna della fisica.* "Il Nuovo Cimento", 4, 162–170 (1927)
121. F. Enriques, *Il determinismo e la fisica quantistica nel congresso fiorentino della "Mathesis".* "Periodico di Matematiche", X, 65–70 (1930)
122. O. Neurath, R. Carnap, H. Hahn, *Wissenschaftliche Weltauffassung* (A. Wolf, Vienna, 1929)
123. F. Enriques, *Philosophie scientifique, in Actes du Congrès International de Philosophie scientifique (Paris, 1935)*, vol. I, *Philosophie scientifique et empirisme logiques* (Hermann, Paris, 1936) pp. 23–37.
124. M. Stoelzner, in *Federigo Enriques e l'Enciclopedia Neurathiana.* "Rivista di storia della filosofia", 3, 483 (1998)

Part I
The Bologna Lectures

Albert Einstein's Lectures in Bologna. Words of Presentation

Federigo Enriques

At the invitation of a university committee for the popularization of the new scientific doctrines, Albert Einstein gave three lectures at the University of Bologna, on October 22, 24 and 26, in front of a very large audience that, not frightened by the abstruseness of the explanations, listened to the Master's lectures with religious attention until the very end, thus demonstrating the fascination and almost the charm of his great personality. Einstein received other invitations from scientific and academic bodies, but—due to numerous commitments—he could only accept the invitation of the Academy of Padua, put forward by the president, Prof. Gregorio Ricci, staying in Padua to repeat, in brief, the explanations given in Bologna.

At the first lecture held in the great hall of the Bolognese Archiginnasio, seat of the ancient Studio, Einstein was introduced by Prof. Enriques, president of the inviting University Committee, with the following words.

Albert Einstein does not need to be introduced to you; fame, which is—in this case—the herald of glory, has already shouted his name; his abstruse doctrines, among the most abstruse to which human thought can rise, are today attracting the attention of the whole world, and not only of mathematicians or philosophers, but of the general public, as can be seen from the place that the author has taken in the communications and even in the polemics of the daily press. This interest, so widespread and so extraordinary in a speculation so remote from the interests of daily life, is worthy of some reflection. In an age clouded by passions, in which the pure ideal of knowledge seems to yield in the face of the exacerbation of appetites and the unleashing of violence, what is the meaning of this emotion aroused by Einstein's theories?

Yet its eminently pacific conquest does not bring the promise of any great utility to alleviate the present crisis of nations, and still less does it offer favorable prospects

---

Federigo Enriques, *Parole di presentazione,*
"Rivista filosofica", II, 1921, pp. 271–274.

      43
S. Linguerri and R. Simili, *Albert Einstein—Italian Memories*, History of Physics,
https://doi.org/10.1007/978-3-031-52950-4_2

to those who approach it eager for profit. If someone asks him "what is it for?" Einstein might answer him with the words which tradition puts into the mouth of Euclid addressing a slave: give him a coin and get him out of here, for he wants to profit from geometry.

In fact Einstein's doctrine appears, today, as far from practice as the objects to which it refers are infinitely large and, on the other hand, infinitely small with respect to the common experience. His mechanics brings to the mechanics of Galilei and Newton a correction relevant only for the great velocities, which approach 300 thousand km to $1''$ that is the speed of propagation of light. The influence of this correction in the study of astronomical velocities is hardly felt on elements that depend on secular terms. Its most remarkable effect is to explain the precession of Mercury's perihelion which rises—in a century—to $43''$. It means that the point at which the planet Mercury is nearest to the Sun will be seen by an observer to be slightly shifted with respect to the prediction given by Newton's theory, in other words shifted by an angle that is about one fortieth of that according to which we see the diameter of the Moon. The extreme smallness of this error (in a hundred years of observation!) makes one think of the marvelous precision of the measurements of the astronomers who have succeeded in pointing it out, and lets us judge how far the Newtonian theory of the movement of the planets, that Einstein proposed to correct, goes in terms of exactness. Nevertheless Einstein is presented to the public as a revolutionary. His doctrine or discovery has brought a new opportunity to cry out that science is bankrupt. Many have rejoiced or grieved that even the firmest truth which for two centuries we have learned to revere as the triumph of human reason, that is to say, Newton's law of universal gravitation, must now be acknowledged to be inaccurate. Not exact, in other words false, however small the error may be to perceivable observation, because reason cannot admit half a term to the alternative of true or false: to be or not to be!

He who judges in this way is far removed, not only from Einstein's thought, but from the historical concept of science now accepted by the contemporary mind and especially by mathematical thinkers. For no theory today pretends to absolute exactness, but each is given as a perfectible degree of truth, which develops and grows with the progress of reason. Thus Einstein's theory does not signify the death of Newton's theory, but rather the conquest of a truer truth, in the face of which the previous one will always appear as a degree of approximation.

To have surpassed this degree, up to the point of explaining the slightest perturbations just hinted at, to have thus discovered the correcting law of barely perceptible errors, constitutes the most splendid triumph of human reason! In spite of all the sophistry with which attempts have been made to distort its meaning, this is also the real reason for the emotion aroused by Albert Einstein. He restores our faith in reason, precisely in this dark hour when it seems to be submerged in the clash of dark passions. He invites us to turn away from the romantic dream of the ego intoxicated by the lordship of the universe, to turn to the contemplation of the order that the mind is able to discover outside itself, in the marvelous work of art of nature. There is already in this invitation a high moral significance. But something higher springs from a close examination of Einstein's thought.

You have already heard that by his criticism he subverts the common concepts of space, time, and motion. To the Kantian philosophers who, in the name of reason, ask one to accept certain judgments a priori, so that science may be possible, he replies that reason has no limits necessarily marked out, that it does not offer a pre-established order for experimental data, but that it finds in itself the power to enlarge the frameworks in which the familiar experience of things close at hand is composed, adapting itself to a more extensive experience. Certainly there is something surprising and almost fearful in this progress of thought which overcomes the limits of its own intuition and fashions higher intuitive forms for itself. How often does the vertigo of flight seem to sweep us into the abyss of the absurd!

Yet this progress presents itself as the logical consequence of a critique which—in seeking to harmonize apparently contradictory data—holds fast to principles and uses them as the key to analyzing the meaning of concepts. In this respect, the philosophical revolution that Einstein brought to completion proves to be the result of an evolution of thought, several centuries old. It began 500 years before the Common Era, with Parmenides of Elea, the first proponent of the relativity of motion. Einstein does not diminish himself by saying that he encloses in a larger cosmological synthesis the work of a long series of philosophers, mathematicians and physicists, from whom he has gathered disparate elements in order to fuse them in his construction.

Let us now listen reverently to the word of the Master—with that reverence which is due to free spirits, and which wishes to be a free discussion of ideas. The doctrine which has already accounted for phenomena confirmed by observation, must be proved by a broader test of experiment; from which a further theoretical development may also result. The author desires this test and awaits it, confident in reason. Unwavering in the face of the sarcasm of easy mockers, he fearlessly faces the paradoxical consequences of his logic. In the same way his ancient predecessor, Parmenides, pursued the ideal of divine truth outside the "opinions of mortals who err far from the true faith", and for this faith Plato - in the *Theaetetetus*—portrayed him in the words of Homer:

august and terrible in his grandeur.

Lecture I: The Special Principle of Relativity

Albert Einstein

The Theory of Relativity arose from necessities deriving directly or indirectly from experiment.

The scientific and mathematical form hitherto possessed to express the laws of Physics was no longer sufficient for the purpose, especially following the development of Electrodynamics and Optics by Maxwell, Hertz and Lorentz.

We know, up to a certain point, the law of propagation of light and think that light itself propagates in a straight line with constant speed c (300,000 km at $1''$) in vacuum. Now, according to Maxwell's ideas on Electrodynamics and Optics, this constant speed c, is independent of the motion of the source emitting the light.[1] This simple law brought the need to transform our ideas about space and time.

Another general truth had been known for a long time: Galileo's law of inertia, according to which a body that is not subject to the action of other bodies, remains in a state of rest or uniform rectilinear motion. For the exact interpretation of this law and the previous one, it is necessary to fix a reference body. This can be practically a rigid body, but for the mathematical treatment it is useful to assume what we call "a system of coordinates" (three indefinite lines coming out from a point and forming two right

---

[1] This constancy and independence is implicit in wave theory (think of the analogy with sound), and in this sense appears as an a priori postulate pertaining to the conception of an immobile aether. Now the a posteriori verification of this postulate is provided in the first place by the astronomical observation of double stars. If the speed of light were to be added together with the speed of the source which now approaches and now recedes from the earth, irregularities would result in the Keplerian motion of the companion around the principal star, which in reality have not been ascertained. Apart from other astronomical confirmations (and from what can be deduced from the experience of Michelson and Morley) there is then a direct verification provided by Majorana's experiment (1917) (*Editor's note*).

Albert Einstein, I1 *Principio speciale di relatività*,
"Periodico di Matematiche", s. IV, vol. II, 1922, pp. 125–135, edited by G. Todesco.

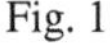
Fig. 1

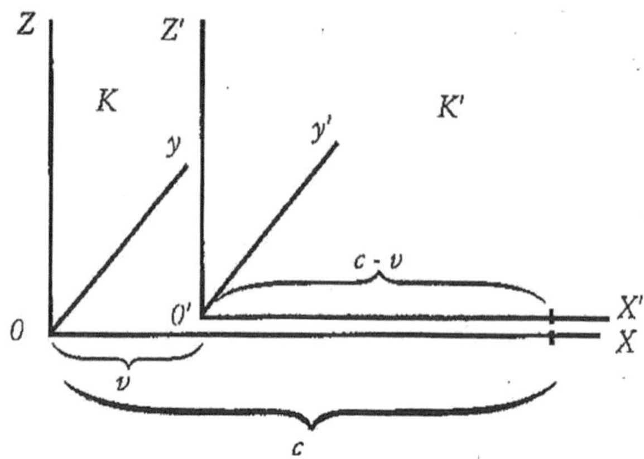

angles). Then the law of inertia can be expressed as follows: "There exists a system K of coordinates in such a state of motion that, relative to it, the body considered travels a straight line with constant velocity". Let us now imagine a coordinate system K' initially coincident with the system K and animated with respect to it, of a uniform rectilinear motion. It will be legitimate then to affirm that the body considered also moves with uniform velocity with respect to K' (it moves in a straight line with constant velocity). That is to say: in accordance with the law of inertia, the systems K and K' (and so all the infinite other animate systems of uniform velocity with respect to K) are equivalent. So that, generalizing, we can say:

> If K is a coordinated system moving with respect to K' uniformly and without rotation, then natural facts take place with respect to K' with precisely the same laws as with respect to K.

This result is called the "Principle of Relativity" (in the special sense).

Since this principle is valid for all the laws of Classical Mechanics, one may wonder if it is of universal validity. A little reflection seems to show that the principle of relativity is not applicable to the law of propagation of light. Let K and K' (Fig. 1) be in fact two coordinated systems. Let K' be animated with respect to K by a uniform rectilinear motion at velocity v. Let a ray of light propagate in the system K along the *x-axis* with velocity c. At the beginning of the times the two systems K and K' coincide. After a second, the extremity of the ray of light, judged by the system K, will have travelled a distance c equal to its velocity; judged instead by the system K' (which in the meantime has progressed with respect to K by a distance v) it will have travelled a space c-v. Therefore, in accordance with the law of propagation of light, the two systems K and K' are no longer equivalent and the principle of relativity does not seem to be able to be applied to this law, because the expression of the speed of light referred to the system K is different from that of the same speed referred to the system K'.

It was then decided for the non general validity of the principle of relativity and it was consequently admitted that there should exist a privileged system of coordinates (K_0) in particular conditions of motion, which because of this privilege could be considered as at absolute rest (with respect to a hypothetical luminous aether), while all the other systems K, mobile with respect to K_0, would have been in motion with respect to said aether. For this system K_0 would have been valid for the law of the constant speed of propagation of light, for the others not. Except that, having admitted this, it was instinctive to think of an experiment capable of highlighting the motion of the co-ordinate systems K (and in particular the motion of the earth's translation) with respect to the aether. This experiment should have proved, in substance, the influence of the translatory motion of the Earth on the velocity of the propagation of light: that is to say the anisotropy of the physical space of the Earth in the different directions. In other words, a ray of light should have propagated with different velocities, along a direction parallel to the translatory motion of the Earth and along a direction perpendicular to it.

Messrs. Michelson and Morley set up an experiment of this kind on the details of which we do not wish to linger here and obtained a negative result in the sense that they could not prove the different speed of light propagation in the two aforesaid directions. They verified instead that a system of instruments designed to highlight a certain physical phenomenon always provides with respect to the phenomenon itself the same results however it is oriented with respect to the translational motion of the Earth.

There is therefore no agreement between the previous theoretical reflections and experiment and it seems that the principle of relativity in a special sense is also applicable to the law of propagation of light, contrary to what we said before.

To explain the contradiction it has been tried, with various hypotheses, to account for the impossibility in which a terrestrial observer finds himself, of noticing, by means of experiments performed on Earth, its motion with regard to the aether. Traveling in the opposite direction instead, the Theory of Relativity (that we will call "special" to distinguish it from an amplification of it that we will examine later), has succeeded in putting the principle of relativity in a narrow sense in agreement with the law of propagation of light, blending them together in a logical construction not in contrast with any experiment.

According to this theory, the principle of relativity in the narrow sense was admitted as absolutely valid and the question was asked:

What form must the laws of Nature take so that, with respect to their expression, all coordinate systems in uniform relative translatory motion are equivalent and the law of the constant speed of propagation of light in different media is nevertheless respected?

In order to answer this question we take again into examination: (a) the Principle of Relativity in a narrow sense; (b) the law of propagation of light in vacuum and we ask ourselves if there is real disagreement between them. Reflection shows that it does not and points out that the arguments previously put forward in order to detect the disagreement itself, were not right in the sense that they contained some assumptions that were unnecessary and due only to habit.

The most important of these assumptions concerns the concept of contemporaneity. Before the advent of the special theory of relativity this concept was believed

Fig. 2

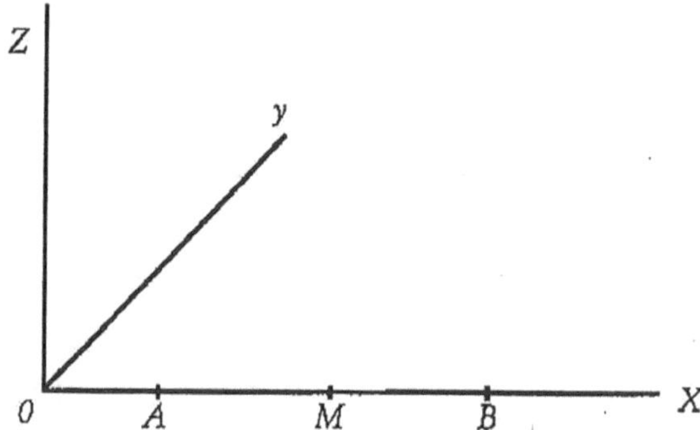

to be self-evident and did not need to be defined. The phrase: "Two events are simul-taneous" was thought to be so intuitively comprehensible that no further investigation of its precise meaning was required. But careful reflection shows that such *a priori* evidence for the concept of contemporaneity does not exist. Suppose, in fact, that two individuals disagree in asserting that two phenomena have occurred simultaneously. They can argue all they want: they will never be able to convince each other. This happens because, since there is no definition of contemporaneity, there is no basis on which to profitably base the discussion. So, wanting to give any definition of contem-poraneity and remembering that our difficulties on the subject we are interested in come from the law of propagation of light, let us start from this law to enunciate that definition.

Let us take a system K of reference coordinates such that Galileo's law of inertia applies to it (inertial system) and suppose that in two points of it A, B (Fig. 2), far enough away, two lightning strikes fall. We look for a means to establish if they fell at the same time or not. Let us join the two aforesaid points with a line and take the middle point M, of the segment AB so identified. Let an observer placed at M be provided with an optical apparatus such that he can view the points A and B together. It is clear that under these conditions the contemporaneity of the two mentioned events can be established, asserting that: "The two events are contemporaneous if the observer placed in M perceives at the same instant the luminous sensations coming to him from A and from B, otherwise they are not contemporaneous".

So the contemporaneity of two events with respect to the same body of reference is defined in a simple way and it is obvious that this definition, supposing nothing about the nature of light, is not in contrast with the law of the constant speed of propagation of this latter in the different directions. It is now necessary to apply the given definition to the case of inertial systems in relative motion. Let K' be an inertial system in uniform translational motion with respect to K. The lightning has still struck the points A and B of the system K, and we have established, by means of the above definition, the contemporaneity of the two events with respect to the observer located in M. We may ask: For an observer located in the system K', will the two events still be contemporaneous? A simple reflection shows no. If the law of the constant speed of propagation of light is valid for the two systems, the contemporaneity of the

Fig. 3

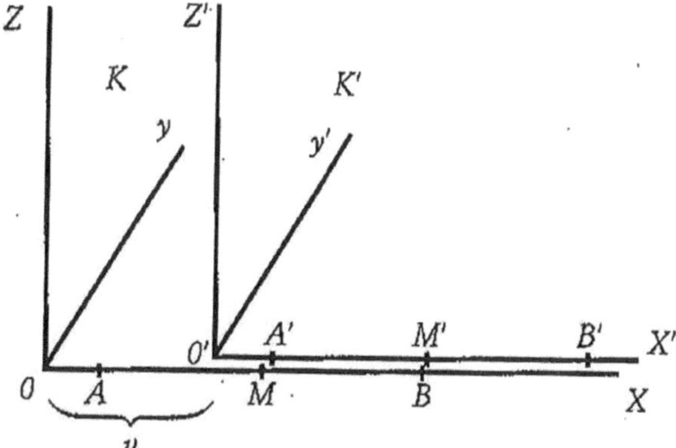

two events must be relative. In fact, to the two places AB of the system K in which the two events fall, correspond two places $A'B'$ (Fig. 3) of the system K', and to the midpoint M of AB, the midpoint M' of $A'B'$. As long as K and K' are both at rest, the point M' coincides with the point M.[2] That is, two events that are contemporaneous with respect to K are also contemporaneous with respect to K'. But if K' is animated with respect to K by uniform translatory motion, then it happens that the two rays of light departing from A and from B (which, meeting at the same instant in M, give the notion of contemporaneity with respect to K of the two events in question) no longer also meet in M' because this point, which is part of the system K', has moved in the time unit of a line v equal to the velocity with which K' is animated with respect to K. That is, the two events are no longer contemporaneous with respect to the system K' and, in general, with respect to any other inertial system that moves with respect to K with uniform translational motion. This simple reasoning leads us therefore to assert that each body of reference has its particular contemporaneity.

And then it is understood that the preceding arguments (which had led us to assert that the principle of relativity in a narrow sense was not applicable to the law of propagation of light), because they were founded on the false assumption that the notion of contemporaneity was independent of the reference system, were wrong. Introducing the concept of relativity of the contemporaneity and that of the relativity of the spatial distance that we will immediately examine, the relativistic principle in a narrow sense is applicable also to the law of propagation of light in conformity with the experimental results obtained by Michelson and Morley; and the contradiction of which we previously spoke disappears.

Let us examine again in fact the reasoning made to demonstrate the non validity of the relativistic principle in a restricted sense with respect to the law of propagation of light. We had said: the extremity of a ray of light that propagates along OCX in the system K, travels in one second the distance c, and this is right; on the other hand the system K' travels with respect to K in the same unit of time the distance v, and

---

[2] In this reasoning and in other analogous ones, we disregard the fact that the two systems K and K', because of the impenetrability of the bodies, would continually obstruct each other.

this is also right. So that the extremity of the ray of light, judged by K', will have traveled in one second the distance c-v instead of the distance c. Here lies the error. In fact the time interval of one second with respect to the system K cannot be equal, because of the above considerations, to the time interval of one second with respect to the system K', therefore it is false to admit that the extremity of the ray of light travels along the distance c-v in a time that, judged by K, is equal to one second.[3]

The second assumption, deriving from habit, consists in believing that the shape of bodies is independent of their motion. For example, it was believed until now that the length of a segment belonging to an inertial system K' in uniform relative translational motion with respect to another inertial system K, was the same, both if judged from K' and if judged from K. Now this assumption is not a necessary point. That is, it is not said *a priori* that by measuring the length of the segment on K' and repeating the measurement on K, the two measurements must give equal results (Relativity of the notion of spatial distance). So if the ray of light travels in the unit of time, measured on K', the segment c-v, this, measured on K, may not be equal to c-v.[4]

In order to develop further the theory that (abandoning the two hypotheses of the independence of the temporal and spatial distance of two events from the state of motion of the reference system) succeeds in putting the principle of relativity in a restricted sense into agreement with the law of propagation of light, it is necessary to find certain relationships that allow us to determine the place and time of an event with respect to K, knowing the place and time of the event itself with respect to K'; in such a way, however, that the law of the constant velocity of propagation of light is satisfied for both systems. These relationships were found by Lorentz and their application allows us to pass from an inertial system K to any inertial system K' endowed with uniform relative translational motion with respect to it. It is thus found that given two inertial systems K and K' of which the latter is in uniform relative translatory motion with respect to the former, the clocks located in K' judged by K, mark the seconds more slowly than it appears judging them by K'; while the lengths located on K' measured by K appear shortened. This second result finds its confirmation in the negative outcome of the aforementioned experiment by Michelson and Morley.

---

[3] Here it is convenient to point out that the measurement of time, for every place, that is to say with regard to a given reference system, supposes a postulate of agreement of the clocks, whose content is analysed in Enriques "Problemi della Scienza", chap. IV. Electromagnetic physics explains the reason for this agreement by referring to a hypothesis on the constitution of the atom: the atom, consisting of an ion (positive) around which some electrons (negative) revolve, provides the natural clock, to which all the ways of measuring time are reduced, since each electron describes its revolution around the centre in a constant time that can be taken as a unit, just as the periods of revolution of the planets around the sun (year, etc.) provide the system of astronomical measures of duration, for our solar system.

[4] That the relative character of contemporaneity brings as a consequence the relativity of length measurements, is seen as follows. If a stationary observer wants to measure the length of a moving train, he must consider two positions of the ends A, B of the train that are contemporary *for him*; but these positions are not contemporary for the observer who is on the train, who finds a different measure. Further examining the question, we recognise that the first observer (fixed) finds a length less than that which appears to the second observer (mobile with the train).

In fact, in order to explain this result, they were forced to admit by way of hypothesis that the bodies in motion underwent a shortening in the direction of the motion (Lorentz contraction), a shortening that is instead, as we have seen, simply explained by the theory of relativity, without the need to make it the subject of a subsidiary and particular hypothesis. But there is more. If the principle of relativity is valid for all the laws of nature, these must remain unchanged passing from an inertial system to another by means of a Lorentz transformation. This is expressed by saying that the laws of nature are covariant with respect to the Lorentz transformation.

This principle provides us with the mathematical means to prove the correctness of the theory of relativity. If even a single law of nature were found that was not covariant with respect to the Lorentz transformation, the theory itself would fall inexorably. But up to now all the research done in Electrodynamics and in Optics has been such as to confirm such covariance. Also the laws of Classical Mechanics become covariant with opportune modifications that we refrain from taking into examination. The need for such modifications is avoided when we remain in the context of very small velocities in comparison to that of light, as is the case for almost all terrestrial velocities, but it becomes evident when we must consider speeds of order of magnitude comparable to that of light. This latter is however an unreachable limit velocity.

A very important consequence of the special theory of relativity concerns the nature of mass and energy. The material energy of a point of mass m is expressed, according to this theory, by the formula[5]:

$$E = m \frac{c^2}{\sqrt{1 - \frac{v^2}{c^2}}}$$

from which we find that the energy becomes infinite for $v = c$, that is to say when the body is animated by a velocity equal to the speed of light, which represents, as we said, a limit condition that can never be reached. Developing in series the preceding expression, we have:

$$E = mc^2 + m \frac{v^2}{2} + \frac{3}{8} m \frac{v^4}{c^2} + \cdots$$

from which we see that, leaving out the first term in which the velocity does not appear, and limiting ourselves to the 2nd term, we have an expression of the kinetic energy that coincides with that known by Classical Mechanics. The following terms instead contain the higher powers of order equal to the velocity v and are ordinarily

---

[5] In reality we add here a hypothesis that exceeds the pure relativistic view, that is that all the forces are of electric origin and subject to Maxwell's equations. In truth, in order to evaluate E the definition of the force intervenes, that is measured with static experiments but that it is necessary to transform in the passage from a reference system at rest to a system in motion. This passage is performed through Maxwell's formulas. See Einstein, "Jahrbuch der Radioactivität und der Electronik", January 22, 1908.

neglected for their smallness. Let us suppose now that we have a box containing many balls all equal (representing for example the molecules of a gas). The inertia of the whole complex, according to Classical Mechanics, is independent of the motions by which the balls may be animated with respect to the box that contains them. That is to say, according to this concept, the same force is always necessary to accelerate the motion of the box both if the balls are still and if they move with respect to it. According to the special theory of relativity instead, the inertia of the box depends on the velocity that may be possessed by the balls contained in it. So, if the balls acquire a certain kinetic energy, the inertia of the box increases: that is to say, the inert mass of a body, according to the new concepts, is a function of the energy it possesses. The two theorems of the conservation of mass and energy are thus identified in a single theorem until the body acquires or loses energy.

Lecture II: General Relativity

Albert Einstein

We want first of all to summarize briefly the concepts inherent to the Special Theory of Relativity, developed in the past lecture. We have seen how the said theory originates from the difficulty met by researchers in order to bring into agreement two laws deriving from experiment, whose validity is put beyond doubt: (1) the principle of Relativity in a narrow sense, that establishes the equivalence of all the inertial reference systems in relative uniform translatory motion, in accordance with the description of the laws of nature; (2) the law of the constant speed of propagation of light in different directions, independently of the motion of the light source. At first sight it seems impossible to bring the two laws into agreement, but it does not take long to realize that this impossibility is only apparent and disappears when some modification is introduced to the ideas that dominate classical kinematics. And precisely that it is necessary to accept, for the bodies in motion, Lorentz's contraction and the delay of the clocks connected to them. To develop the theory of Special Relativity, it is necessary to require that the laws of nature remain unchanged with respect to a Lorentz transformation. It can then be said that the motion of bodies is only relative, at least the uniform translatory motion, because all the states of motion of this type that can be imagined are equivalent in the above sense.

We can now ask ourselves whether this equivalence does not exist, also, from the physical point of view, for non-inertial systems or, which is the same thing, for states of non-uniform motion. This equivalence evidently exists from the kinematic point of view, but from the physical point of view, that is to say to the effects of the description of natural laws, nothing can be said a priori about it. Indeed, it seems, on a first superficial examination of the question, that it is not legitimate to consider non-inertial systems (that is to say those systems for which Galileo's law of inertia is not valid) as equivalent for the aforesaid purpose: the very law of inertia just

Albert Einstein, *La relatività generale*,
"Periodico di Matematiche"", s. IV, vol. II, 1922, pp. 221–231, edited by G. Todesco.

recalled opposes it. Let us assume in fact an inertial reference system K. In it, as far as we know, the bodies move in a straight line with uniform velocity. Then let us consider a second system S_A that moves with respect to the first in accelerated motion with constant acceleration. The bodies moving with respect to K that describe a straight line with constant velocity instead describe, as we know, a parabola with respect to S_A. It is no longer legitimate to speak of equivalence between the systems K and S_A with regard to the effects of the motion of these bodies, because Galileo's fundamental law is not applicable to the system S_A. It is possible however to apply to the accelerated systems a more general law than Galileo's, with respect to which the principle of the equivalence of all the inertial and accelerated reference systems can easily be sustained. To this aim we recall a theorem that, known since the times of Galileo, was never correctly interpreted by the experts of Classical Mechanics. We mean to refer to the theorem of the equality between the inertial mass and the gravitational mass of a body. First of all we notice that, given a body, its mass can be defined in two completely different ways:

(1) as *inertial mass*, that is, as the ratio between the accelerating force imparted to the body and the corresponding acceleration it receives, which is usually expressed by the formula

$$Inertial\ mass = \frac{Force}{Acceleration}$$

In this sense two or more bodies have different masses if, applying the same accelerating force to them, they receive different accelerations.

(2) as *gravitational mass*, that is as the ratio between the force acting on the body when it is immersed in a gravitational field and the intensity of the field itself, which is commonly expressed by the formula:

$$Gravitational\ mass = \frac{Force}{Field\ strength}$$

In this sense two or more bodies have different masses if, when placed in the same gravitational field of given intensity, they are stimulated by different forces.

The two definitions of mass thus established are totally different. However, experiment does not allow us to detect any quantitative difference between the two masses: inertial and gravitational. Now, from the preceding relationships we derive:

$$Gravitational\ mass = \frac{Gravitational\ mass}{Inertial\ mass} \times Field\ strength$$

But experiment tells us that, in a given gravitational field, all bodies acquire the same acceleration, independently from their physical nature, so the ratio between the two masses, must be equal for all the bodies. This ratio can be reduced to unity with a convenient choice of the units of measurement, and then it follows that:

Inertial mass = Gravitational mass

A result so elementary and precise, must naturally have a profound cause, but classical mechanics has not been able to grasp it completely.[1] We will see at once what importance the right interpretation of this result has for our study, but first we wish to show a weak side of Galileo's law. It says: A body situated far enough from other bodies not to be affected by their action, remains in a state of rest or of uniform rectilinear motion. Now it is not possible to ascertain whether a body is or is not subjected to the action of other bodies except by examining its motion with respect to them. Suppose we have a system S_A that moves with accelerated motion with respect to an inertial system S. The bodies that are at rest with respect to S appear to be endowed with the same acceleration with respect to S_A, and from the relativistic point of view this can be interpreted in two ways: either by saying that the said bodies are at rest and S_A is in accelerated motion, or by saying that S_A is at rest, but the bodies contained in it enjoy the property of falling with equal acceleration if they are left to themselves. This is equivalent to postulating the existence, outside the system S_A, of a gravitational field, since the above property is, as far as we know, suitable to characterize such a field. In other words: the fact of knowing that, in a gravitational field all the bodies, falling, acquire the same acceleration, gives us the right to treat the system S_A as if it were at rest and to think that a gravitational field exists, outside it. It is then understood how Galileo's law is not applicable to these systems. This law in fact is valid only for bodies very distant from each other, such that the reciprocal effects of the relative gravitational fields can be considered null. It will be convenient therefore to continue in the idea just expounded. If it is true that every accelerated system can be replaced by a gravitational field, it will be possible by calculation to study all the properties of the field itself, deducing them from those of the accelerated system with a simple change of coordinates. In other words, if it is legitimate to consider the accelerated systems as at rest in the above sense, then it is possible to construct a general theory of gravitation.[2] A first application of the above is obtained by examining the trajectory of a light ray passing through a gravitational field. From what has been said, this trajectory, which appears rectilinear with respect to an inertial system, must appear curved with respect to an accelerated system. This theoretical deduction is subject to experimental confirmation referring to a very intense gravitational field such as can be, for example, for us, the one

---

[1] However the reason for this appears clear if we admit the metaphysical hypothesis underlying the speculations of Galileo and Newton on the atomic structure of matter: atoms being heavy (according to Democritus) and qualitatively identical, mass and weight would be proportional to the number (or to the volume) of the atoms contained in a body (*Editor's note*).

[2] This is true at least in first approximation. By extending the limits of space and time the equivalence would be lost; under these conditions an accelerated system is equivalent only to particular gravitational fields. Further examining the question in relation to the following developments of the article, it is recognized that a general field of gravitation can be substituted for the accelerated motion of a reference system in the same way as any metric determination of a multidimensional variety admits at each point a tangent Euclidean metric determination.

Fig. 1

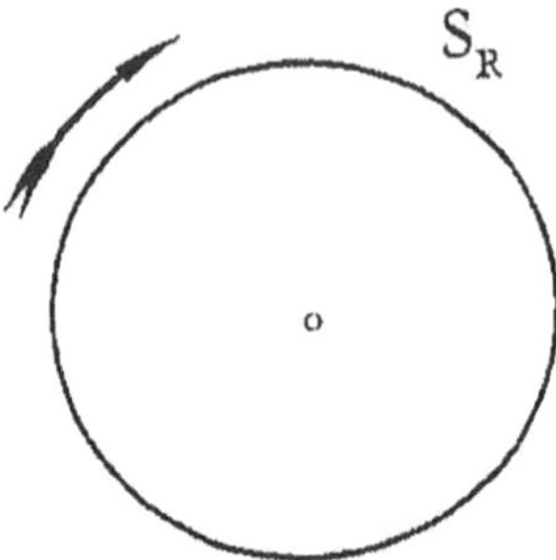

inherent to the solar mass. We will see better later on how experiment has furnished this confirmation in the most unexpected way, during the solar eclipse of 1919.

Having arrived at this point of our study, it would seem easy to construct the general theory of Relativity. It would be enough to study the natural phenomena bringing into play, for the accelerated systems, consideration of the gravitational field. But the matter is not as simple as it appears at first sight and instead it is complicated in an extraordinary way. This derives from the fact that, wishing to make the notion of gravitational field intervene in the study of natural phenomena, that is to say wanting to consider reference systems in any condition of motion, it is necessary to abandon the limited field of Euclidean geometry and to attribute a new and more general meaning to the space–time concepts.

Let us take as reference system S_R a disk rotating with uniform velocity around its center C (Fig. 1).

For this system Euclidean geometry is no longer valid and it is no longer possible to define time as was done in the previous lecture. Let us first ask ourselves what physical meaning the geometrical ideas of space and time have. Referring to a Galilean system K, we see that bodies are so made that they can be set in motion as Euclidean geometry requires. In practice this means that for the Galilean system K (Euclidean) it is possible, for example, to make the following construction. Let us take a large number of small rods, all equal, and let us arrange them in such a way as to form a network of squares. The construction is fully possible and it is such that the net entirely covers any surface belonging to K. But suppose now that the small rods have the property of expanding with temperature and that the temperature changes from point to point of the surface considered. Our net of squares will be in disarray because the various small rods, dilating in various ways according to their position on the surface, will be deformed in such a way that the squares constructed in the aforesaid way will no longer close exactly as before and the net will no longer be such as to cover the whole surface. Wanting however to consider the rods as if they were equally long, it is no longer possible to apply the procedures of Euclidean geometry to the surface under examination. If we substitute for the temperature any other cause the reasoning does not change and so it is possible to realize that the space in which we operate is no longer Euclidean. Such is the case of our rotating disk S_R. In fact it is known that a body in motion undergoes the "Lorentz contraction" because of

Fig. 2

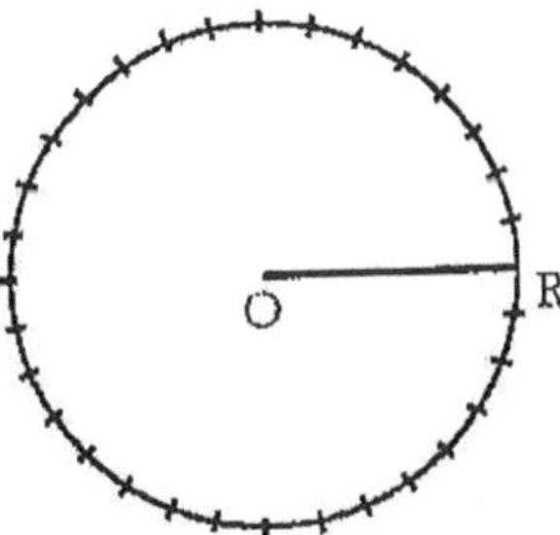

which it appears shorter than it really is. Now let us arrange many small rods along the radius *OR* of the circumference *C* of our disk (Fig. 2); and let us ask ourselves if Euclidean geometry is valid for it. It is immediately seen that the small rods placed along the circumference, since they are disposed in the direction of the motion, will undergo a shortening, while those placed along the radius, since they have a direction perpendicular to the motion, will remain unchanged. Therefore the ratio between the circumference and the diameter of the disk will no longer be expressed by π, but by a number greater than π.[3]

Here the theorems of Euclidean geometry are no longer valid for this reference system, as had been postulated. It should be noted here that the theorems of geometry have meaning in Physics if it is possible to apply them to real objects that come within the scope of our senses. Otherwise it remains only an independent logical construction. Now the disk S_R can be taken as a reference system in the physical sense, and since at its periphery a (centrifugal) force acts for which the theorem of the equality between the gravitational and inertial masses is valid, it can be said that with respect to it a gravitational field exists. But, since Euclidean geometry is no longer valid for this reference system, it is not possible to fix a system of coordinates nor to speak of space in the physical sense. Analogously, if we take two clocks *A*, *B*, (Fig. 3) and place one at the periphery and the other at the center of the disk, since the first one has a certain speed and the second one does not, it follows, for a known result, that the first clock, judged by a Galilean system, marks the seconds more slowly than the other. That is to say, the gravitational field has an influence on the behavior of metric rods and clocks. As far as time coordinates are concerned, an experimental confirmation can be deduced from the above. The gravitational field created by the disk in motion has a radial disposition and the clocks arranged along a line of force must, as we have seen, run all the more slowly the farther they are from the center of the disk. This must be true also for the gravitational field created by the stars, which has the same radial disposition. Then if we consider two electrons whose vibrations furnish a certain monochromatic light, placed one on the surface of the Sun and the other on that of the Earth, the vibratory motion of the first must

---

[3] This statement would require an in depth analysis. To penetrate more deeply into the question it is useful to consider a fixed circular platform and a closed train running along the periphery; the length of the train must be measured by an observer standing on the platform, and he must consider the head and the tail of the train in a position that is simultaneous *for him.*

Fig. 3

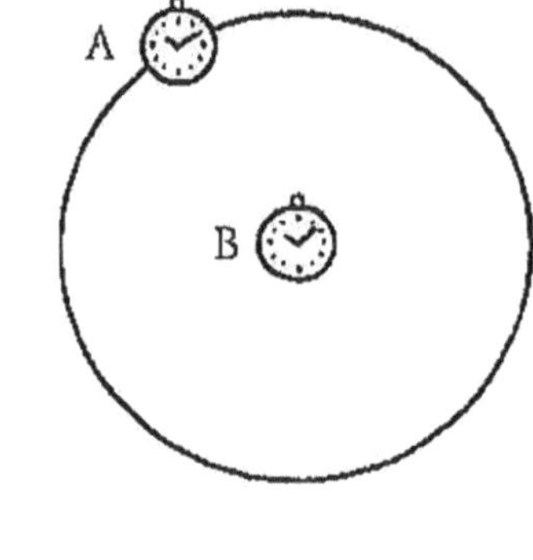

Fig. 4

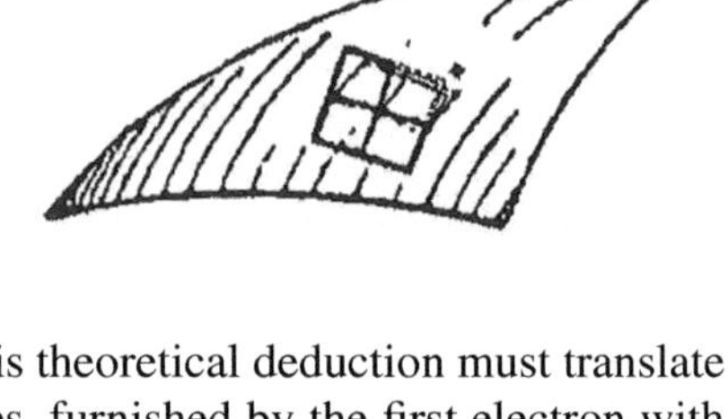

appear slower than that of the second one. This theoretical deduction must translate experimentally into a shift of the spectral lines, furnished by the first electron with respect to those furnished by the second one, toward the less refrangible region of the spectrum. Unfortunately, there are various causes of error that overlap to mask the phenomenon, so that the experiment in question could not yet give decisive results[4]. If the experiment succeeds, it would be possible to go back to the expression of the potential of the gravitational field of the stars and to calculate the mass of the stars.

The difficulty encountered, for non inertial systems, of not being able to give a sense that is directly connected with experience to the space–time coordinates, can be happily overcome by means of certain considerations that we are going to expound. The non-Euclidean continuums were especially studied by Gauss, who indeed furnished the theory relative to two-dimensional fields. To expound it briefly, let us refer to a portion of an ellipsoidal surface (Fig. 4).

In the meantime we immediately see that we are in a non-Euclidean field because it is impossible to construct the network of squares of which we spoke previously on the said surface. Gauss then considers a double system of numbered curves on the surface such that the curves of the same system have no point in common (Fig. 5), while each curve of the first system meets each curve of the second in one point and in one point only. In this way, for each point of the surface one curve and one curve only of each system passes and the point itself is fully identified by assigning the two numbers p, q (Gauss coordinates) relative to the curves that meet in it. With this system it is possible to identify all the points of the surface considered, without knowing anything about their reciprocal distance. To be more precise: the

---

[4] Recent experiments confirming the theory have been made by Pérot (Cf. the *Comptes rendus de l'Institut de France*, March 7, 1921).

Fig. 5

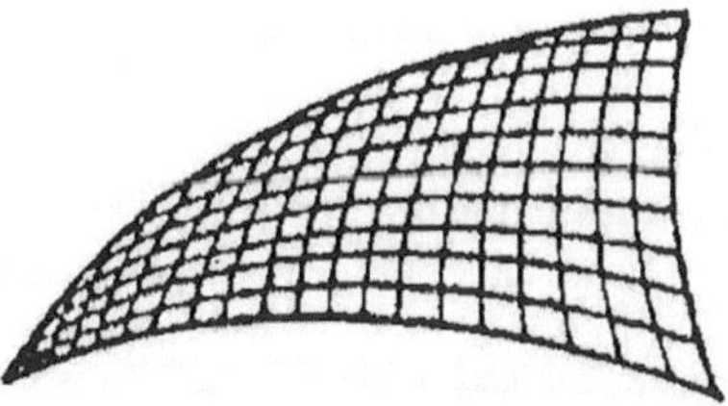

distance between two points can be determined only as a function of the coordinates themselves ($ds^2 = g_{11}dx_1^2 + \ldots$ where $g_{11}, g_{12}, \ldots$ are functions of the coordinates). If these functions take on certain simple values the surface becomes Euclidean and the reference system becomes Cartesian. Moreover, the system of Gauss coordinates (that has been extended to continuums of three and more dimensions) is applicable only to those continuums that can, within small enough portions, be treated as Euclidean. We then proceed by approximation by examining how the expression of the coordinates varies as the portion of the surface considered increases. Now, we have seen that the existence of a gravitational field ensures that the necessary methods of measurement are no longer, in the general case, those given by Euclidean geometry.

However, as far as the Earth's gravitational field is concerned, its influence is too small for us to realize the non validity of Euclidean geometry. We can however reach the same result by reflection. We already know that it is possible to replace the consideration of an accelerated system by that of a system at rest if we introduce the existence, outside of it, of an adequate gravitational field. So if we think of an observer falling vertically in a gravitational field being endowed with the same acceleration as the field, it is clear that for him there is no gravitation. If in fact it lets bodies fall around him, they will remain stationary with respect to him because they will immediately acquire the same acceleration of the field, with which he himself is endowed. Precisely it is seen that at every moment, there exists around each point of space a four-dimensional neighborhood as the result of which the geometry and the motion of the clocks is such as it would be if the gravitational field did not exist. In other words, considering gravitational fields of infinitely small dimensions, the system of co-ordinates relative to them comes to acquire an immediate physical meaning, in the sense of special relativity, because it is possible to choose the state of acceleration of the local co-ordinate system to be infinitely small so that the existence of the gravitational field has no perceptible influence.[5]

This means that, operating in portions of space infinitely small with respect to the vastness of the universe of which we are part, it is possible for us to consider Euclidean geometry valid in first approximation, despite the presence of our gravitational field. For the totality of space, the four space–time coordinates have no physical meaning, but they acquire it instead if we consider spaces and times infinitely small with respect to the universal four-dimensional continuum.

---

[5] We mean to say that within sufficiently narrow limits special relativity is valid, so that the distinction between space and time acquires within these limits a physical meaning that is missing instead for those who take into consideration the four-dimensional universe that is space–time.

The problem inherent in non-inertial systems reduces to a mathematical problem: the determination of those *g*-functions which appear in the expression of the distance *ds* considered above. They provide the analytic expression of the gravitational field and also give the geometric qualities of space and time inherent in it.[6]

From this perspective, the gravitational field and the metric field are identified: Geometry and Kinematics on the one side, gravitation on the other, since they provide the metric properties of four-dimensional space, are equivalent. And since the former give the concept of inertial mass and the latter that of gravitational mass, a physical interpretation of the theorem of the equality between the two inertial and gravitational masses is acquired in the simplest way.

The considerations above show how it is possible to extend to the accelerated or non-inertial systems the results of the special theory of Relativity and to arrive at the knowledge of the general law of the gravitational field. This law must be such as to remain invariant for any transformation (and not only for a Lorentz transformation, as in special relativity). Having obtained the general field law, it is important to note that it provides in first approximation Newton's classical law. It is therefore a broader law, not a new law. In second approximation, the law in question accounts for certain anomalies that Newton's law alone could not explain. For example, the precession of the perihelion of the orbit of the planet Mercury. According to Newton's law, the planets must move along orbits whose orientation, in comparison to the fixed stars, must remain unchanged over the course of the time. Now for a long time it had been observed that the orbit of the planet Mercury moved slightly with respect to the said stars. Various explanations of this anomaly were attempted, but the theory of general relativity accounts for it in a perfect and satisfactory way. If the calculation is made starting from the general law of the gravitational field, instead of from Newton's law, it is seen that the orbits of the planets must change position over the course of time, in very slight measure, so slight that the shift is observable only for the planet Mercury, which, being as is known nearer to the Sun than the others, feels the influence of the gravitational field of the Star in a more intense way.

---

[6] For every mathematical expression of ds^2 (subject to certain inequalities) a force field remains defined in general, since every geodesic of space–time represents the trajectory of a motion that can be provoked by the field. However, the physically simplest and most interesting case is that of fields generated by isolated rather than diffuse masses. The treatment of this case leads to a system of differential equations which the g coefficients of ds^2 must satisfy. The physical meaning of these conditions does not yet seem sufficiently clarified.

Lecture III: The Relativistic Conception of the Universe

Albert Einstein

We expounded in the last lecture the ideas that inform the theory of General Relativity. We want now to dwell particularly on a consequence of the theory itself, liable to experimental confirmation. Then we will summarize all the results obtained and we will go on to expound the relativistic conception of the Universe.

We have seen that the trajectory of a light ray must curve in a gravitational field. To verify experimentally this conclusion it was necessary to operate in a very intense gravitational field. Having to exclude the field inherent to the stars because the angle under which they are seen by a terrestrial observer is too small to be easily appreciated, because of their enormous distance from us, it was thought to take advantage of the gravitational field inherent to the solar mass. And precisely to observe the trajectory covered by a light ray coming from a star opposite to the Sun and skimming the solar disk along its path. According to the theoretical predictions, the light ray should deviate from the original rectilinear trajectory and bend towards the Sun, undergoing a measured deflection of $\frac{1.75''}{\Delta}$, Δ being the distance of the trajectory from the center of the Sun, expressed in radii of the Sun itself. How could such a deflection be practically observed?

Let S and S_1 (Fig. 1) be two stars opposite to the Sun. A luminous ray coming from S should propagate according to the rectilinear trajectory SA, but because of its passage in the immediate proximity of the Sun it is affected by the action of the gravitational field due to the mass of the Sun and undergoes a deflection approaching the edge of the solar disk, so that its trajectory becomes hyperbolic and the ray itself propagates along the branch of hyperbola SB. So that an observer located on the surface of the Earth T, sees the star in the direction C, and precisely in S'. In an analogous way it will seem to him that the light coming from the star S_1 emanates from S_1'.

Albert Einstein, *La concezione relativistica dell'Universo*,
"Periodico di Matematiche", s. IV, vol. II, 1922, pp. 231–236, edited by G. Todesco.

Fig. 1

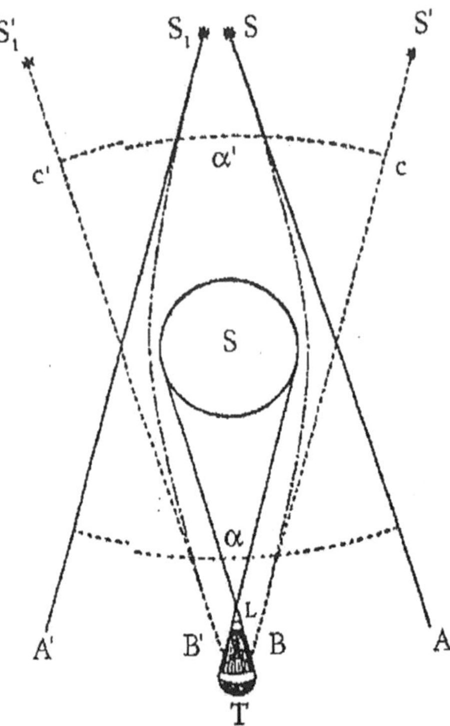

Then the angle formed by the directions of the light rays coming from the two stars S and S_1, will increase because of the presence of the solar mass and assume the value $\alpha' > \alpha$. From the experimental point of view, we only have to take two photographic images of the two stars S and S_1, the first one at a time when the Sun is interposed between the same stars (which can be done only during a total eclipse of the Sun, because otherwise the light emanated by it, diffused by the atmosphere, would make the images of the stars in question evanescent), and the other one some time later, that is when the Sun occupies another position in the Sky. If our arguments are right, it must be possible to verify that in the first photograph the distance between the images of the two stars is perceptibly greater than what appears from the examination of the second photograph. The astronomers Eddington, Crommelin and others, in fact photographed 7 stars opposite the Sun during the total eclipse of May 29, 1919 in Sobral (Brazil) and in Principe Island (West Africa). Appropriately compared with each other, the copies obtained with those relating to photographs taken earlier, when the Sun occupied a different position in the sky, showed a difference in the distances of the stellar images of about 1/20 of mm, in agreement with theoretical predictions within approximation limits of 10/20%.

Another confirmation of the phenomenon in question may occur on the occasion of the total eclipse of the Sun that will occur in the coming year.

Let us now briefly summarize the course of ideas followed in our exposition. We treated first of all the special relativity that assumes as a hypothesis the validity of Euclidean geometry for fixed bodies of reference. In this hypothesis the coordinates possess a physical meaning of measurement that allowed us to identify the behavior of clocks and metric rods. The foundation of this first part of the theory is the law of propagation of light with constant velocity, relative to a given system of coordinates. This law derives from experiment and from the study of the Electrodynamics of moving bodies (Maxwell, Lorentz, etc.), but it seems to be in contrast with the principle of special relativity. The contrast is however only apparent and it is sufficient to modify the kinematic ideas of universality, of time and of the laws relative to bodies and clocks in motion, to realize that the two principles do not contain anything contradictory, and can combine very well in a completely logical theory. Abandoning the prejudices inherent to this false universality brings with it the necessity to admit the Lorentz contraction and the delay of the clocks for bodies in motion, and leads to the concept of relativity of contemporaneity. At the heart of the whole theory now summarized is the fact that natural laws are covariant with respect to the Lorentz transformation, which allows us to give mathematical expression to the Principle of Special Relativity. In the above, however, we do not find an explanation of gravity nor a correct interpretation of the theorem of the equality of the two inert and gravitational masses of a body. We are therefore forced to widen the field of validity of the Special Principle of Relativity and to put forward the hypothesis that systems of non-inertial coordinates are also equivalent to the effects of the description of natural laws. In this order of ideas it must be possible to determine in what sense gravity acts on the bodies in motion and to what extent it exerts its action on physical laws, namely to give the general law of gravitation. Once the expression of this general law is obtained, it is found that it furnishes in first approximation of Newton's well known law; in second approximation, it gives the reason of the anomaly found in the motion of the planet Mercury and of other anomalies not sufficiently explained until now in the light of classical theories.

As long as one remains in the field of Classical Mechanics and the Special Theory of Relativity, geometry exists as a discipline in itself independently from physics, that is to say, the way of admitting the existence of physical bodies is completely independent of the metric properties of space. But when one moves on to consider the General Theory of Relativity, this independence no longer exists and is undermined by the presence of the gravitational field that modifies the geometric structure of space. So, for example, it has been seen that in the case of a disk rotating uniformly around its center, Euclidean geometry cannot be kept valid anymore. Now if the existence of a gravitational field brings with it the non validity of Euclidean geometry, we can ask ourselves how the space that constitutes our Universe must be perceived, which cannot be considered Euclidean anymore as the mechanics of Newton would require.

However, we could think that the space that surrounds us differs very little from a Euclidean space, that is to say, it was "quasi-Euclidean". Let us see what the theory of Relativity suggests in this regard.

Until the age of Riemann, geometers had put forward the hypothesis of a spherical space, i.e., having a finite volume, without being limited by this, and this hypothesis had always remained in the purely geometric field. Let us see how it is possible today also to attribute to this concept a physical meaning. What do we know about the disposition of matter in space? Since the relative speed of the so-called fixed stars is very small in comparison to that of light, it is legitimate to think in first approximation that the matter existing in the Universe is immovable. In first approximation it is also legitimate to admit that the density of this matter is constant, that is to say that the same matter is uniformly distributed in space.

This being admitted, it is shown by calculation that the ideal case now presented gives as a solution a spherical world whose radius depends on the density of the matter constituting it, according to the formula

$$R = \sqrt{\frac{2}{K\rho}}$$

where ρ is the average density of the matter and the ratio $\frac{2}{K}$ has the value 1.08×10^{27} in the C.G.S. system. Now for $\rho = 0$ we have $R = \infty$ and therefore the world would be infinite and unlimited, as Newton's theory requires, provided that the average density of matter was zero. In this hypothesis, the Universe would admit a kind of center in which the density of matter would be maximum and, moving away from which, it would decrease until it would cancel itself completely at infinity. This representation of the Universe is, in our opinion, not very satisfactory. Admitting that the average density ρ of matter is different from 0, then we have for R a finite value and the world can be spherical, finite and unlimited, in the above sense.

Indeed, however, matter is not uniformly distributed and is not immobile as has been assumed in approximation, so the concept of the spherical world will only be approximate. In reality, it will be a quasi-spherical world, but still finite, because the equations from the gravitational field show that wherever matter exists, the curvature of space is positive. Is it possible, in the present state of our knowledge, to acquire a perceptible idea of a finite but unlimited world? This question can be answered by beginning to consider a spherical space in two dimensions (spherical surface): let us make a stereographic projection of the sphere on a plane, and let us consider, above the sphere, the magnitude of its plane projection as the magnitude of an arc; then the sphere—from which the center of projection is removed—appears precisely as a finite but nevertheless unlimited surface.

And it is clear how similarly we can proceed considering a space with three or more dimensions. And so a perceptible idea of a finite and unlimited world, as ours could be, is acquired on the basis of the results, that are drawn from the examination of the relativistic theories expounded so far.

Part II
Correspondence

Albert Einstein-Federigo Enriques Correspondence

F. Enriques to H. Zangger

Bologna viale Gozzadini 9, April 20, 1920

Dear Professor Zangger

On my return from a trip to Egypt I received your kind letter. I hasten to reply that I have no new editions of my "Problemi della Scienza" in view. My later studies, broadened in the social sense, are contained in the volume "Scienza e razionalismo" published by Zanichelli in 1913. Now for some years (compatibly with my work in Mathematics) I have immersed myself in research to reconstruct the earliest history of Greek science.

With best regards, your devoted servant

Federigo Enriques

I enclose a letter for Einstein.

To Professor Einstein

Dear and illustrious Colleague

Prof. Zangger's postcard, which also bears your signature, gives me the welcome opportunity to express my heartfelt gratitude for the many proofs of interest in my studies on the theory of Science, which you have shown me over the years, and in particular to reply to a postcard which I received at the beginning of the unfortunate war, which for that reason I was unable to reply to.

The consent and esteem of a superior spirit such as yours constitute one of the most sought-after rewards to which my work could aspire. Not as a mere exchange of courtesy, but as an expression of sincere feeling, I will tell you that I have followed your brilliant scientific career from the beginning: already in 1904 in Paris some eminent friends (such as Perrin) spoke of you with the greatest enthusiasm. Of course I am now informed of what no one is unaware, that is, of your latest great success, with which you succeeded in making an essential step in Newtonian theory: nevertheless I could not say that I have assimilated well the spirit of your guiding ideas, in which I am convinced there is something more than mathematical formulae.

S. Linguerri and R. Simili, *Albert Einstein—Italian Memories*, History of Physics, https://doi.org/10.1007/978-3-031-52950-4_6

It is my fervent wish to meet with you under conditions conducive to relaxed conversation. Who knows when I will be able to fulfill it? In the meantime, please accept, dear Colleague, the expression of my warm and devoted feelings…

yours, Federigo Enriques

F. Enriques to A. Einstein

Bologna viale Gozzadini 9, 19 January 1921

Distinguished Colleague,

An expressly constituted committee wishes to promote annual conferences of distinguished foreign scientists at our University of Bologna. It is desired to begin the matter in the highest manner by inviting here a superior personality to give lectures or, possibly, discussions on an object of universally recognized importance. No name has seemed equal to yours and no subject so exciting to the scientific world as relativity. I come, therefore, to express to you the earnest desire of my colleagues to have you here to inaugurate our scientific conferences in Bologna. You yourself, if you are in the main willing to accept the invitation, will judge how many lectures may be sufficient to give the learned public some idea of your doctrines and discoveries, and whether it would be advisable to devote a special session to a more restricted conversation in which you may give to the more qualified those clarifications or explanations that they may desire.

I must now touch upon another point. The Committee, which is at the beginning of its work, does not have large sums of money which can worthily compensate your science and your fame. We can only offer a travel and subsistence allowance, which we could calculate overall at 3000 Italian lire. And we wish to flatter ourselves that the modesty of the offer will be compensated by the desire to reciprocate the sympathy with which your studies are generally followed in Italy, and especially in this most ancient University. To which we would like to invite, for the occasion, our colleagues from the sister Universities.

I will be grateful to you, dear and illustrious professor, if—by answering this letter, as I hope, favorably—you will also tell me if it is possible for you to explain yourself in Italian or French. If not, it would be necessary to organize things, for example by publishing your lectures in German in advance and preparing an interpreter. But of course the impression on the audience will be much stronger if you can, in whole or in part, explain yourself in Italian (or French).

Having thus expressed the wish of our Committee to you, I need not add that I personally would be glad to see you here and to make your personal acquaintance, and to attest to you by word of mouth the sentiments of gratitude for the sympathy that you have shown to my philosophical work, sentiments that I had already expressed to you in a recent letter.

Consider me meanwhile, with admiration, devotedly yours

Federigo Enriques

If you are willing to come we can easily change the date, e.g. around the Easter holidays, in the season when a trip to Italy is most welcome, or at a time that suits you best.[1]

F. Enriques to A. Einstein

Bologna viale Gozzadini 9, February 17, 1921

Dearest Colleague,

As I was on my way to Rome, I was late in receiving your very kind and welcome letter of 2 February. I am very happy that you are in principle willing to come: we will now discuss the timing. I am also very happy that you can express yourself in Italian, which will make your words rewarding. You explain yourself very well and the few grammatical errors that you may make are of no importance. I will say more: I can see that you are familiar with our language, or that you were so in the past, at the age when languages were learned. Now, experience has taught me an axiom: languages are not forgotten, because if the occasion arises you find them again. So it happened to me with French, and I thought it was not possible: I had gone to Belgium and I returned from there speaking French fluently, as in my childhood. So it will happen to you for Italian.

Let us talk now about the time of your coming among us. In September the University here is deserted; it begins to repopulate towards the middle of October; and - if you are not going to America—you could well come in the second fortnight of that month. But it seems to me that next June would be a better season, and preferably one of the first weeks of June itself. Then the professors and students enjoy the greatest freedom, while waiting for examinations to begin; and I would add that - usually - there is also favorable weather for traveling in Italy: it is not yet excessively hot even at midday. Think about it and get back to me. With best regards to you.

F. Enriques
(Headed Società Italiana di Matematiche "Mathesis")

F. Enriques to A. Einstein

Bologna viale Gozzadini 9, April 25, 1921

Dear Distinguished Colleague,

I don't know if this will reach you while you are probably still in America. I received your kind letter of 17 March, and I learned with pleasure that you will be able to come here in the second half of October. According to your wishes we shall find an interpreter who will listen to your speech and who will perhaps be ready to translate what you are able to express only in German. I shall write to you again about this before the summer holidays.

---

[1] *Footnote written in Enriques' own handwriting.*

In the meantime I am glad to inform you that our Academy of Sciences of Bologna wished to express its affection and admiration for you by electing you as its correspondent.

With kindest regards, yours

Federigo Enriques
(Headed "Periodico di Matematiche")

F. Enriques to A. Einstein

Bologna viale Gozzadini 9, 11 July 1921

Dear and illustrious Colleague,

On the occasion of your visit, a University Committee was formed here, under the honorary chairmanship of the Rector of the University, composed of Prof. Ciamician, Prof. Perozzi, Prof. Flora, Prof. Puntoni, Prof. Tarozzi, Prof. Maiocchi, Prof. Valenti and Count Cavazza, as well as of the president here undersigned; and on behalf of this Committee (which represents all the Faculties of our University) I extend to you my greetings and thanks in advance.

I have found the person who will be able to act as interpreter, helping—if necessary—in your exposition. He is Prof. Bianchi, who teaches German language and literature at our University and who is very familiar with the language because he has spent long periods in Germany. Now some Colleagues observe that the audience would be better prepared to understand your lectures if you would send them in advance a written summary, which you could write in German and translate into Italian and publish, distributing it to the listeners. If you would like to gratify this wish, I would be grateful if you would send me these summaries as soon as possible, since printing is not done here very quickly.

For the time of your coming among us, we will remain, as agreed, for the end of October. If we also go to the beginning of November, the University will be fuller. I also expect that some colleagues from other universities will come here for the occasion. It would be good to make a more precise program, and to make it known soon.

With the most devoted and cordial sentiments, your

Federigo Enriques
(Headed "Periodico di Matematiche")

F. Enriques to A. Einstein

Bologna viale Gozzadini 9, September 12, 1921

Dearest Colleague,

We are approaching the time for your lectures, promised for the second half of October. Now I need to know, soon if you will be so kind, the exact days on which you propose to give these lectures, and preferably also their titles.

I myself must be absent at the end of this month, going to Naples for a conference of our mathematical society, of which I am president, which—for various reasons—has had to be fixed for the 13th, 14th, 15th and 16th of October; but I will return here directly, so that you may freely fix the beginning of your lectures, if you like, from the 20th or even the 18th of October.

Now before I leave Bologna, because of the said absence, I would like to arrange everything that pertains to your lectures, both for the hall, for the invitations, etc., and so on. That is why I would like a somewhat precise program from you. If then you can favor us, besides the titles of the lectures, with summaries of them too (as I already mentioned in a previous letter before the summer) it would be very useful. If you can hint at this, I will arrange for these summaries to be received here during my absence and, if they arrive in time, translated and organized for the benefit of the listeners. But, I repeat, what I need now is above all the precise program of your lectures. I would add that it would not be inconvenient if these lectures were to be extended until the beginning of November.

With best regards, yours

Federigo Enriques
(Headed "Periodico di Matematiche")

F. Enriques to A. Einstein

Bologna viale Gozzadini 9, September 25, 1921

Dear Distinguished Colleague,

I receive with great interest your kind letter, from which I have the confirmation of your coming to Bologna, in the second half of next month.

Following this, your three lectures are fixed for Saturday, October 22, Monday, October 24 and Wednesday, October 26.

In this way, as you kindly hint, and as I am sure is the wish of your most scientific listeners, there will be the opportunity to arrange some meetings en petit comité: since many of us are anxious to have very many explanations from you, as you can well imagine. At the same time you will be able to leave from here on the 28th for Florence, and stay in that city on the 29th and 30th, as you indicated is your wish.

I will be back in Bologna, from Naples, on the 18th or 19th at the latest, and if you arrive here on the 20th (October), please let me know, so that I can have the pleasure of seeing you at once.

In any case, I beg you to come to my house for tea on Friday evening, the 21st, at 9 o'clock (that is to say 9 p.m.): I hope to introduce you to some colleagues and friends before the lecture that will take place the following day at 3 o'clock (that is to say 3 p.m.). It is understood that your son, who is just the same age as mine, will come with you.

As far as the hotel is concerned, you can stay at the Hotel Pellegrino, or at the Savoia. If you want more luxurious hotels, there are the Baglioni and the Brun. But also in the other more modest ones, and especially in the Hotel Pellegrino, I think that you will be fine. I think it is better to book in advance.

Now, as I will be leaving Bologna in a few days, I have made arrangements for you to receive an answer and clarification on what you may need. Therefore, even during this period, you can address letters to me here in Bologna, which will be opened by my wife, and of which I will also be informed immediately.

In conclusion, I await from you only the most precise communication about your arrival here, counting on seeing you at my house at the latest on the evening of

October 21; and rejoicing to make your personal acquaintance soon, I send you in the meantime the most cordial greetings. Yours

Federigo Enriques

F. Enriques, P. Langevin, P. Weiss, P. Montel, G. Malfitano to A. Einstein

Paris, 30 December 1921

Gathered together in a place where you came on another occasion, we are happy to offer you our most affectionate admiration.

F. Enriques,
P. Langevin,
Pierre Weiss,
Paul Montel,
G. Malfitano

F. Enriques to A. Einstein

Via Sardegna 50
Rome, 8 February 1923

Dear Colleague,

I am writing to you from Rome where I transferred to last year.

I remember that when you came to Bologna I had occasion to speak with you of the desire that many people in Italy would have of having you here permanently among us, something that would be a real stroke of luck for our Italian University. But you amicably explained to me the reasons why it would not be convenient for you to leave Berlin.

Now they say that conditions in that city have changed and that—for reasons of anti-Semitism—you are no longer comfortable there and are about to leave that place and Germany too. If this is the case, the hope is reborn that we can somehow win you back to our country. This idea and this desire is to be found in many and is only waiting for an opportunity and encouragement to manifest itself and take concrete form. I limited myself to talking about it with the Minister of Education, who is the idealist philosopher Prof. Gentile, and he authorized me—albeit in strict confidence—to tell you that he is for his part willing to welcome an initiative in this regard.

I thought it opportune to explain to the Minister how you, in your situation, have reason to desire above all a great deal of freedom, and he understood the matter perfectly and told me that—if you come round to the concept of accepting a position in Italy—he is willing to study a way of satisfying you. And for my part I add that, for this purpose, one could seek in the meantime an opportunity to have you come here for some lectures and have the opportunity to discuss the practicalities of the thing verbally.

In the meantime, please have the goodness, as soon as you receive this letter (which I do not know how to address except to your old address in Berlin), to reply in one line, which I will hasten to communicate to the Minister. I do not need to ask you to consider this letter as <u>confidential</u> in the meantime, as the Minister has expressly asked me to prevent the press from getting hold of the idea.

I am grateful to take this opportunity to remind you of me, recalling the unforgettable days in Bologna, with devoted friendship.

Federigo Enriques

A. Einstein to F. Enriques

Berlin W. 30, 11/4/1923 Haberlandstr(asse) 5

Dear colleague!

I was much moved by your letter, and I must confess to you openly that I would prefer your company, and that of Levi–Civita, to that of my colleagues here. But I suffer no harm from anti-Semitism, although it is present in a very high degree. It has instead the effect that I am left more in peace than I would be otherwise. In addition, I am very closely connected to my present environment by family, friendship and business ties. At my age it is not so easy to change milieu, because you no longer have enough elasticity to be able to blend into a new one.

For all these reasons, with every feeling of gratitude and affection for you and for your country, which is always particularly dear to me, I cannot make up my mind at this moment to accept your kind proposal. But if in the future I should feel obliged to leave this nest of mine, because of an aggravation of the situation, I will turn to you at once with joy and confidence.

Hoping to see you again soon, I am with affectionate regards to you and to Levi–Civita, your

A. Einstein

F. Enriques to A. Einstein

Via Sardegna 50
Rome, 18 November 1925

Dearest Colleague,

The Treccani Institute that is attending to the publishing of an Italian Encyclopedia must have written to you inviting you to collaborate with an article on the Theory of Relativity.

Now, as Director of the Mathematics and Mechanics Section of the Encyclopedia, and also on behalf of all my colleagues on the Steering Committee, I now ask you to accept this invitation.

The Italian Encyclopedia, which has taken as its model the Encyclopaedia Britannica, seeks to harmonize the popularizing character of culture with a highly scientific character. It will be deemed an honor to have an article written by the very author of the greatest physical synthesis that has been achieved; and for many of us this name sounds not only as illustrious, but above all as that of a dear friend.

I dare therefore to hope that you will accept my prayer and send me, in the meantime, an acceptance in principle that will allow your name to be announced among the contributors; and then we can agree on the limits within which the article would remain and which other entries should be referred to.

I would add that, as to the fee, you yourself are invited to say—as you may do, especially to me, in complete freedom and confidence—what sum should be paid to you.

In the meantime, consider me, with friendship, your affection and most devoted...

Federigo Enriques

P.S.—In Milan, in the last few days, I have seen our mutual friend Langevin, who came there, like me, for the Congress of the Federated Intellectual Unions. This is a movement that cannot fail to have your support; and I hope that, as it spreads—as it seems—in Germany, you too will give your name to it; and also that there will be occasions for us to meet.

F. Enriques to A. Einstein

Via Sardegna 50
Rome, 14 April 1926

Dear Colleague,

My proposal for an international bibliographical work on the history of science has been followed up, as you can see from the report of a meeting held for this purpose at the Paris Institute. It has now been arranged that a new small meeting of experts will soon have to be held in Paris to flesh out the proposal. A larger international committee will then be formed to sponsor the enterprise.

I have been instructed to make enquiries as to the persons who might most suitably undertake to take part both in the said committee and in the more restricted meeting of experts, referred to above. And I would therefore appreciate your valuable advice.

It would be especially important to indicate a technician who, in addition to a certain authoritativeness, also has the practical skills that are most necessary to arrive at a concrete work proposal.

You yourself may be embarrassed to suggest a name for this purpose, but it will not be difficult for you to obtain the necessary information. It seems to me that what is needed is a man who has some experience of libraries and who, for his part, would immediately find out what has been done in Germany or what support could be relied on: I mean work support, that is, the appointment of suitable people.

I would be very grateful if you would reply to this letter as soon as possible because there is not much time to prepare experts.

Renewing to you the expression of the pleasure I had in meeting you lately in Paris, with friendship.

Your most devoted

F. Enriques

F. Enriques to A. Einstein

Via Sardegna 50
Rome, 20 February 1930

Dear Distinguished Friend,

The University of La Plata addresses to me a request for suggestions for a chair of mathematics with the office of head of the mathematical department. The same request was addressed to you, and to Professors Borel and Langevin.

I immediately thought of Professor Alfred Rosenblatt of the University of Cracow, whom I had already proposed two years ago, with Levi–Civita's support, for the Faculty of Buenos Aires. Rosenblatt is a distinguished mathematician, whom I know well because—at first—he pursued research in the direction of the Italian geometric school, on algebraic surfaces. After this early and very profound work, he occupied himself with questions of Mechanics and Hydrodynamics, so as to earn much esteem from Levi–Civita. But he was never able to settle down in his homeland (and he must have been between 40 and 45 years old) especially for reasons of anti-Semitism and because of some different interests of members of the Faculty of Cracow.

Under these conditions it would be very desirable that Rosenblatt's proposal could succeed. And, from the objective point of view of the Faculty of La Plata, I believe that it would be difficult to make a better choice: seriousness of research, extensive scientific activity in various areas of mathematics, are combined in Rosenblatt.

I therefore take the liberty of recommending the matter to you warmly. I know that Messrs. A. G. Doranona and Enrique Gaviola of the University of La Plata are here in Berlin, although I have not been given their address. You will probably have occasion to confer with these gentlemen, whose influence may be decisive.

I am grateful to reaffirm with devoted friendship, your

F. Enriques

Ready to give details of Ros[enblatt]'s scientific activity where necessary.

Someone even in Berlin must know his work or at least some of his work (maybe Lichtenstein).

A. Einstein to F. Enriques

27 February 1930
Professor Dr. Federigo Enriques
Via Sardegna 50
Rome

Dear Colleague!

I have a warm feeling for your and Levi–Civita's motives and will recommend Mr. Rosenblatt as prime candidate, however I too would have an excellent candidate to propose (Dr. Walter Mayer of Vienna) who is in a similar situation. I do this considering that splitting our votes must be avoided as the worst thing.

I am with cordial greetings
Yours

A. Einstein

F. Enriques to A. Einstein

Rome via Sardegna 50, 15 April 1930

Dear Colleague,

I had your kind reply in the past about Rosenblatt and the mathematical chair at La Plata, and I am most grateful to you for it. R. has written to me that he saw you in Berlin and also attests to his gratitude for your great kindness. Let us hope that he will be able to obtain the post!

Following the death of my friend E. Rignano, I return to the editorship of "Scientia" which I had founded with him 25 years [ago].

The Journal needs to establish itself more and more and therefore it needs the support and collaboration of the greatest. I would like to be able to count on your support, and not only for the courtesy you have always shown me. The initiative of a Journal of an international character like ours, I believe, must be congenial to you.

On this subject let me speak to you in more friendly confidence. I have the impression that Germany in general remains cold towards the Journal after the war. Perhaps it was regrettable that it abandoned a purely scientific terrain to discuss burning questions that, at the time of the war, it was better to leave aside (and this was my thinking even then, so much so that I left the editorship precisely in 1914).

However, if you really want to attain peace of mind, it is better not to recall, anywhere, what you have said or done during a period that you wish to be over.

I wanted to mention to you—in strict confidence—a doubt or impression about the reception of the Journal in Germany, because I would like to ask you to help us to dispel, if need be, such an impression, as other German friends, including nationalists, have promised to help me for this purpose.

And now, after this long discourse, allow me to ask you for the most effective help that you can give to our Journal: with your work and your name. I know well that the request risks being indiscreet since you do not have much time left for popularizing articles which are in keeping with the character of "Scientia", but I would allow myself to suggest a subject by asking you a question, that is to say what do you think of the indeterminism of Quantum Mechanics. A little article of yours on this subject, even if very short, would, no doubt, be of the greatest interest for the readers.

Can I somehow hope to have your cooperation in the near future?

Yours, with devoted friendship,

Federigo Enriques

Albert Einstein and Other Italian Correspondents

A. Einstein to P. Straneo

Wittelsbacherstr., 13 Berlin-Wilmersdorf.
7 January 1915

Distinguished Colleague!

Together with this letter I send you my latest, concise work on the theory of general relativity and on the theory of gravitation in which this difficult problem has been brought to a certain conclusion. I am convinced that in principle the road taken is the right one and that in the future they will be surprised that the concept of general relativity has found such resistance. Whether motion—uniform or not—can be defined and conceived only as *relative* motion, this character of relativity being its own not only from the kinematic point of view but also from that of general physics.

In these times, sharing so painfully the sentimental ravings of almost all people and their sad consequences, I love science twice as much. It is as if a wicked epidemic had confused people's brains! A fortiori, rational people, and especially we scientists, must cultivate international relations and keep away from the coarse sentiments of the rabble; unfortunately, in this respect even among natural scientists we have had to witness great disappointments!

I have the impression that I met you (about 19 years ago) personally in Italy (Pavia, Casteggio); at that time I was 16 years old. We probably met at the Marangoni family, whose house I frequented a lot.

With best regards, yours

A. Einstein

A. Einstein to T. Levi–Civita

Unspecified location, 17 March 1915

Dear Colleague!

I would be very pleased if next time you would write to me in Italian. As a young man I spent more than six months in Italy; at that time I also had the pleasure of

© The Author(s), under exclusive license to Springer Nature Switzerland AG 2024
S. Linguerri and R. Simili, *Albert Einstein—Italian Memories*, History of Physics,
https://doi.org/10.1007/978-3-031-52950-4_7

visiting the lovely town of Padua and still today I am always pleased to use my modest knowledge of the Italian language. On the other hand I would not have the courage to write to you in Italian, because the result would really be too uncertain and unclear.

I confine myself here to considering your last letter. For at any time we may return to the earlier ones. Your objection culminates in this, that when we diminish the domain Σ in which the $\delta g^{\mu\nu}$ are non-zero, the average $\overline{\delta g}^{\mu\nu}$ values in general do not tend to any limit. As an example you give the function

$$\lambda = \lambda_0 \left\{ 1 - \left(\frac{x - x_0}{h} \right)^2 \right\}^2$$

I will not deal at all with the problem of the existence or non-existence of this limit, because this problem in my opinion is not at all significant; rather I want to try to show that the conclusion drawn from Eq. (71) of the work is correct.[1] We can then connect our subsequent remarks to this demonstration.

Hypothesis

$$\delta I = \int\limits_{\Sigma} d\tau \sum \phi_{\mu\nu} \delta g^{\mu\nu} \tag{1}$$

is an invariant if the $\delta g^{\mu\nu}$ cancel on the boundary of Σ and otherwise can be chosen arbitrarily.

Assertion

$$\phi_{\mu\nu}/\sqrt{-g} \text{ is a covariant tensor.}$$

Demonstration: we have

$$g^{\rho\sigma'} = \sum \frac{\partial x'_\rho \partial x'_\sigma}{\partial x_\mu \partial x_\nu} g^{\mu\nu} \tag{2}$$

so

$$\delta g^{\rho\sigma'} = \sum \frac{\partial x'_\rho \partial x'_\sigma}{\partial x_\mu \partial x_\nu} \delta g^{\mu\nu}. \tag{2a}$$

I multiply (2a) by $\sqrt{-g'}d\tau' = \sqrt{-g}d\tau$ and integrate over Σ. I then obtain:

---

[1] This is A. Einstein, *Die formale Grundlage der allgemeinen Relativitätstheorie*, "Königlich Preussische Akademie der Wissenschaften. Sitzungsberichte", II, 1914, pp. 1030–1085.

$$\int \sqrt{-g'}\delta g^{\rho\sigma'}d\tau' - \int \sqrt{-g}d\tau \sum \frac{\partial x'_\rho}{\partial x_\mu} \frac{\partial x'_\sigma}{\partial x_v} \delta g^{\mu v}. \tag{3}$$

Since the integration domain Σ must be infinitely small, I can substitute the factors

$$\frac{\partial x'_\rho}{\partial x_\mu}, \frac{\partial x'_\sigma}{\partial x_v}$$

with those (constant) values that these factors assume at some point in the integration domain. In doing so, I neglect only something relatively infinitely small in the second member of (3). We have then, if I put for brevity

$$A^{\mu v} = \int_\Sigma \sqrt{-g}\delta g^{\mu v}d\tau, \tag{4}$$

that

$$A^{\rho\sigma'} = \sum \frac{\partial x'_\rho}{\partial x_\mu} \frac{\partial x'_\sigma}{\partial x_v} A^{\mu v} \tag{3a}$$

and this shows meanwhile that $A^{\mu v}$ is a countervariant tensor with components that can be chosen independently of each other.

I now transform the expression of δI. Since $\phi_{\mu v}/\sqrt{-g}$ is a function of coordinates that varies "slowly", I can, neglecting an infinitely small quantity of higher order, consider as constants in the integral of the second member of (1) the quantities $\phi_{\mu v}/\sqrt{-g}$.

$$\delta I = \sum_{\mu v} \left\{ \frac{\Phi_{\mu v}}{\sqrt{-g}} \int_\Sigma \sqrt{-g}\delta g^{\mu v}d\tau \right\}$$

or

$$\delta I = \sum \frac{\Phi_{\mu v}}{\sqrt{-g}} A^{\mu v}d\tau. \tag{1a}$$

Since by hypothesis δI is an invariant and by what was expounded before $A^{\mu v}$ is a countervariant tensor with components that can be chosen arbitrarily and independently from each other, it follows that

$$\frac{\Phi_{\mu v}}{\sqrt{-g}}$$

is a covariant tensor.

With the glad hope that you will not find in this demonstration some essential lacunae and with best regards I am yours.

A. Einstein

A. Einstein to T. Levi–Civita

Unspecified location, 26.III.1915

Dear Colleague,

I have just received your letter of 23 March, written in the Italian familiar to me of which I have been deprived for so long. You cannot imagine how much pleasure it gives me to receive a letter so authentically Italian. Reading it brings back vividly the most marvelous memories of my youth. Even the wording of the letter is very nice: at the beginning you flatter me courteously so that when I read the new objections I do not make a grumpy expression.

First, again, the objection that refers to the special case where by suitable choices of coordinates the $g_{\mu\upsilon}$ are constant: you argue that in this case the $\phi_{\mu\upsilon}$ do not cancel if one chooses the appropriate system such that the $g_{\mu\upsilon}$ are <u>not</u> constant. You have not justified this claim, however, and I do not consider it correct until you have given an example or a general demonstration of it. Let us now turn to the second objection: you admit that

$$\sum_{\mu\upsilon} \frac{1}{\sqrt{-g}}\phi_{\mu\upsilon}A^{\mu\upsilon} + \varepsilon = \text{invariant} \quad \left(A^{\mu\upsilon} = \int \sqrt{-g}\delta g^{\mu\upsilon}d\tau\right)$$

where ε is an infinitely small one of higher order, while you dispute that it can be derived that $\phi_{\mu\upsilon}/\sqrt{-g}$ is a tensor.

I must confess that the motivation for this objection is not clear to me. In fact, since you have shown that the $A^{\mu\upsilon}$ transform countervariantly, it seems to me absolutely irrelevant to the demonstration that these $A^{\mu\upsilon}$ are infinitely small quantities. Nevertheless one must complete the demonstration in such a way that instead of the infinitesimal tensor $A^{\mu\upsilon}$ one introduces a finite one by means of a limit procedure.[2] On this, however, I do not dwell, but I am convinced that you will consider valid the following way of concluding:

Unless you have something relatively infinitely small you have

$$\sum \frac{1}{\sqrt{-g}}\phi_{\mu\upsilon}A^{\mu\upsilon} = \sum \frac{1}{\sqrt{-g'}}\phi'_{\sigma\tau}A^{\sigma\tau'}, \tag{1}$$

for any permissible substitutions, as you yourself agree.

Unless you have something relatively infinitely small—as you still admit—and as I have shown in my last letter, you have

---

[2] In other words, Levi–Civita contests the fact that Einstein, starting from the infinitesimal tensor $A^{\mu\upsilon}$ arrives at determining the tensorial character of the *finite* expression $\phi_{\mu\upsilon}/\sqrt{-g}$.

$$A^{\sigma\tau'} = \sum \frac{\partial x'_\sigma}{\partial x_\mu} \frac{\partial x'_\tau}{\partial x_v} \Lambda^{\mu v}. \tag{2}$$

From (1) and (2) it follows, however, that unless something is relatively infinitely small, this equality also holds

$$\sum_{\mu v} \frac{\Phi_{\mu v}}{\sqrt{-g}} A^{\mu v} = \sum_{\sigma\tau\mu v} \frac{\Phi'_{\mu v}}{\sqrt{-g'}} \frac{\partial x'_\sigma}{\partial x_\mu} \frac{\partial x'_\tau}{\partial x_v} A^{\mu v}.$$

Since the $A^{\mu v}$ can be chosen independently of each other, we have

$$\frac{\Phi_{\mu v}}{\sqrt{-g}} = \sum_{\sigma\tau} \frac{\partial x'_\sigma}{\partial x_\mu} \frac{\partial x'_\tau}{\partial x_v} \frac{\Phi'_{\sigma\tau}}{\sqrt{-g'}}$$

i.e., $\phi_{\mu v}/\sqrt{-g}$ has tensorial character.

It is therefore absolutely unnecessary to introduce some constraint in place of the $A^{\mu v}$, the conclusion being independent of that. Any consideration which would subject the $\delta g^{\mu v}$ to more conditions than are necessary for the nature of the problem is to be ruled out as an unnecessary complication.

I greet you cordially, your

A. Einstein

A. Einstein to T. Levi–Civita

Berlin, 14.4.1915

Dear Colleague,

With your postcard in which you agreed with me you gave me great joy. If there is an opportunity of repeating the demonstration for $\phi_{\mu v}$ I shall introduce the improvements which I have learned through our important correspondence. Meanwhile, to be honest, interest in this subject is absolutely modest. It is strange how little fellow specialists feel the intimate need for a _real_ theory of relativity. Unfortunately, the attitude of the non-scientific community is even incomparably more strange. It is therefore doubly pleasant to get to know someone like you more closely. I will gladly try to make our epistolary acquaintance personal, one more reason for me to return over the Alps. Let us hope that our homelands will not also go to war against each other!

I cordially greet you

A. Einstein

A. Einstein to T. Levi–Civita

Lucerne, 2.VIII.17.

My Dear Colleague!

I take advantage of this holiday-rest in my homeland to send you a *direct* greeting. Recently, I found your beautiful new papers from Hurwitz and I immediately took them away with me to study them here in peace. I admire the elegance of your calculation. It must be nice to ride through these fields on the steed of actual mathematics, while we are forced to make our way on foot.

It gave me particular joy that you welcomed the new term λ, without which it seems to be impossible to form an adequate relativistic picture of the world as a whole.[3] On the other hand, Nature could also have chosen a completely different path; this is only the simplest one conceivable in mathematical terms. Stellar statistics, however, has not so far justified this.[4]

Your objections to my concept of the energy components of the gravitational field still do not persuade me completely. I would like to tell you one more time what makes me insist on my concept. (1) I start from a Galilean space, i.e. one with constants $g_{\mu\upsilon}$. With only the change of the reference system (introduction of an accelerated reference system K$'$) I would then arrive at a gravitational field. If in K$'$ a pendulum clock is placed at rest, driven by a weight, this gravitational energy is transformed into heat, while with respect to the original (Galilean) system K there is certainly no gravitational field, i.e., also the gravitation field energy. Since with respect to K all the components of the problematic "tensor"—energy disappear entirely, if the energy of gravitation were really expressible through a tensor the disappearance of all the components should take place also relatively to K$'$. The example of the pendulum clock argues against this. After all, it seems that the energy components of the gravitational field should depend on the prime derivatives of $g_{\mu\upsilon}$ because this is also true for the *forces* exerted by the fields. But *tensors* of the first order (dependent only on $\frac{\partial g_{\mu\nu}}{\partial x_\sigma}$ & $g_{\mu\nu}$) do not exist.

(2) You think that the field equations

$$G_\mu^\sigma + T_\mu^\sigma = 0,$$

where the G_μ^σ depend only on the first and second derivatives of the $g_{\mu\nu}$, T_μ^σ on matter, are already to be conceived as energy equations, so that the G_μ^σ would be the energy components of the gravitational field. But according to this way of seeing it is totally incomprehensible why in spaces in which gravitation can be disregarded something like a (conservation) principle of energy exists. Why then, for example, should a body not be able to cool itself without emission of heat toward the outside? But the equation

$$G_4^4 + T_4^4 = 0$$

allows T_4^4 to decrease everywhere, while this change is not compensated by a decrease in the absolute value of the quantity G_4^4, which does not decrease in a physically perceptible way.

---

[3] This is the cosmological constant, introduced by Einstein in his 1917 paper.

[4] Einstein is referring here to the discrepancy between the size of the visible universe and the radius of the universe he calculated in the model in footnote 3.

I therefore assert that what you call the energy principle has nothing to do with what in physics is called that. I have formulated a principle in the form

$$\sum_{\sigma} \frac{\partial T_v^{\sigma} + t_v^{\sigma}}{\partial x_{\sigma}} = 0$$

And I have drawn conclusions from this. These conclusions are correct regardless of whether one assumes that t_v^{σ} are "really" energy components of gravitation. Because, with the infinite (spatial) cancellation of the T_v^{σ} and t_v^{σ}, the following principle always applies

$$\frac{d}{dx_4}\left\{\int (T_4^4 + t_4^4)dV\right\} = 0,$$

where the integral is extended to the whole three-dimensional space. All that is required for my conclusions is that T_4^4 is the density of matter, which is not questioned by either of us.

(3) Finally I would still like to draw your attention, appealing to your physical sense, to the following case. Let space be Galilean (free from fields) up to the field that few gravitating masses produce (Newtonian case). According to Newton, the number of force lines branching from them to infinity depends only on the total *mass* of the bodies, according to the results of the theory of special relativity also on the total energy. Now, it seems to me indubitable that, in the static case, the field at infinity must be completely determined *by the totality of the energies of matter and the gravitational field*. This results in my interpretation of the t_v^{σ} as you can easily demonstrate from my field equations in mixed form.

If you wish, I am very willing to elaborate each of these points in more detail than has been done here. I have no literature available here. Best regards.

Yours A. Einstein
Brambergstr. 16A, Lucerne

A. Einstein to R. Academy of Sciences of Bologna

Berlin, 28 June 1921
To the R. Accademia delle Scienze, Bologna

Allow me to express to your Academy my profound gratitude for my appointment as a corresponding member. It will be a great pleasure and a profound experience when in October I see again the Italy I loved so much as a boy. I hope to be able then to participate personally in a session of your Academy.

With the highest esteem

A. Einstein

A. Einstein to R. Accademia Nazionale Dei Lincei

Kiel, 14.VIII.21

To the class of mathematical and natural sciences of the R. Accademia Nazionale dei Lincei

Distinguished Gentlemen!

In your letter of 25 July you informed me of my nomination as a foreign member of your Academy. For this, I express my profound gratitude. This appointment pleases me not only as a sign of esteem and goodwill towards me but also as a happy sign of the re-establishment of friendly relations among scholars and among the scientific bodies of the various states, in the form in which they existed for the good of science before the political catastrophe of 1914.

With the highest esteem

A. Einstein

B. Caldonazzo to A. Einstein

"Scientia"
The Editorial Office
Milan, 25/9/22
Foro Bonaparte, 43
Prof. A. Einstein
Berlin-Wilhersdorf

Sir,

The present state of the new theory of relativity, the degree of maturity to which it has now reached, the developments it has achieved in the large number of publications devoted to this subject, the enthusiasm it has aroused, on the one hand, especially among mathematicians, and, on the other, the doubts and the difficulties of clearly understanding it encountered especially among those who need to see behind the mathematical symbol the physical reality it represents accessible to intuition: all of which makes it seem to us, more than interesting, necessary, that our journal (of which you undoubtedly know the importance it has acquired among the community of scholars throughout the world because of its illustrious collaborations and its enormous circulation, both of them international, and because of its other inquiries into today's most significant scientific questions; we send you one of the latest issues as a gift) to proceed from now on—with that methodicalness and that aspiration to completeness, which have earned its other investigations the highest resonance in intellectual and scientific circles all over the world—to a great international inquiry into this theory of relativity, which will constitute with great probability one of the milestones in the development of science.

An inquiry that must aim above all, and at the same time, at the following two fundamental objectives:

1. Submit the theory TO AN OBJECTIVE CRITICISM, SERENE AND THOROUGH, which highlights the weak points that still need to be revised in order not to give rise to further objections;
2. TO WORK AT CLARIFICATION, trying to render—either the theory as a whole or in its fundamental points, or the principles that constitute its basis or the results at which it arrives—accessible to all scholars who have philosophical interests, AND THEREFORE TO ALL THOSE WHO ARE NOT MATHEMATICIANS.

It is therefore necessary to choose the questions from this point of view, and to adopt an expository method THAT LEAVES NO ROOM FOR DEVELOPMENTS, NOR MUCH LESS FOR MATHEMATICAL FORMULAS. It is necessary that the author should endeavor to present his ideas in a language that is essentially ordinary, and at the same time clear and accessible even to those who, although they have heard much about this theory, have not yet come to grasp its fundamental essence, its principles and its fundamental results, nor, above all, have been able to SEE the physical meaning behind its mathematical structure.

The fact of not being able to make any use of mathematical formulas undoubtedly limits the number and the nature of the questions concerning such a theory of relativity, questions that can be dealt with in our investigation; But there will certainly be those of a more general order and of a more philosophical content—and therefore those more suited to the epistemological character of our journal—which will lend themselves best to be treated with the use of ordinary language and in a manner accessible to scholars of all branches of science; FOR THEM, TOO, AND NOT ONLY FOR SPECIALISTS, THIS INQUIRY MUST BE MADE. Among the questions likely to form part of our inquiry, we think we may mention the following:

The theory of relativity from a physical point of view;
The evolution of physical concepts in the theory of relativity;
The experimental basis of relativity;
The classical postulates of physics in the theory of relativity;
The notions of space and time, in physics, at present;
Times and contemporaneity of relativists;
Gravitation; the interdependence of gravitation and the metric of space;
The gravitational field from a material point;
The equivalence principle in Einsteinian theory;
The inertia of energy;
Einstein's theory under astronomical control;
The propagation of light in space-time;
Electrodynamics in the theory of relativity;
Einstein's theory applied to the world of atoms.

May we therefore hope, my dear sir, that one or other of these questions—as well as others which you may choose at your pleasure, provided they are of a general character and susceptible to being treated without recourse to mathematical symbols—may stimulate you to write the desired article for us? Your illustrious name and the eminent contributions which you have already made in this field of study will certainly make your article one of the most interesting in the whole inquiry.

Each article in the survey should not exceed a MAXIMUM OF 10 printed pages, equivalent to a total of 3700 words; if this is not enough space, you can take twice as much, but then please divide the article into two parts, EACH WITH ITS OWN SUBTITLE, so that they can be published in subsequent issues.

At the end of the investigation the articles will probably be republished in a separate volume that will certainly constitute a unique document, of inestimable value, in the history of science.

We therefore beg you, dear Sir, not to refuse us such a great pleasure. We would be truly and heartily obliged!

Begging you for a kind reply, I express to you, Sir, my sentiments of the highest consideration.

One of the Editorial Secretaries

Bruto Caldonazzo

E. Rignano to A. Einstein (Handwritten message at the bottom of the letter sent to Einstein by Bruto Caldonazzo on September 25, 1922)

Dear Master,

I had already asked you last year, when you gave your very interesting lectures in Bologna, to renew your collaboration with our journal, in which in the past you had already written one of your first articles on relativity. Could not our present inquiry, of which I need not point out the exceptional importance, induce you to satisfy our earnest desire to have the great honor and the great pleasure of counting you again among our most illustrious collaborators? It is precisely with your name that we would like to resume in our columns the German collaboration interrupted by the war; and we truly hope that this may be for you one more reason not to refuse us the great pleasure that we ask of you and for which we would be deeply grateful!

Awaiting your reply, I beg you to accept, Sir, the expression of my sentiments of the highest consideration.

Your Eugenio Rignano

A. Einstein to E. Rignano (Minute of reply written on the back of the letter sent by Bruto Caldonazzo on September 25, 1922)

Unfortunately, I am currently so busy that I cannot write the desired article. Best Regards.

A.E.

P. Straneo to A. Einstein

Genoa, 15/2/1928

Distinguished Professor,

Several years ago, during the war, we were in correspondence and you had the kindness to recall that we had met much earlier, probably in Zurich, where I believe we met with my school friend M. Besso.

Now, after having been an engineer for many years, I returned to my studies and I am professor of mathematical physics in Genoa. I have done a lot of work on relativity and I think I have also sent you a large booklet, as well as various specific works on the integration of gravitational equations.

Now I would like to return to these studies because I believe I can completely solve the problem of integration of the static equations in the case of axial symmetry already taken to a certain point by Weyl, Levi–Civita and myself.

I am also dealing with some quanta theory issues.

So I would very much like to have possibly your work on the two topics from 1916 onwards, since we do not have the Berl. Berichte. Do you still have any extracts available?

I had hoped to see you in Como and we were all disappointed by your absence.

We hope to see you in Bologna at the Congress of Mathematics. Meanwhile please accept, Distinguished Professor, my most respectful greetings.

Your devoted
Paolo Straneo

A. Einstein to P. Straneo

8 May 1928
Mr. Paolo Straneo
Genoa

Dear Colleague,

Due to a long illness I only find time to reply to your kind letter today. I really remember seeing you as a student. I looked with great interest at your papers, in particular the one on the axisymmetric gravitational field. I am sending you some of my own work. I want to direct your attention in particular to the Kaluza theory. According to my belief, in spite of all efforts, the relationship between gravitation and electricity is still an unsolved problem, the most important one in this field. Despite its great achievements I am rather skeptical of quantum mechanics. The renunciation of causality is unbearable for me.

I greet you cordially.

Yours

P.S. I can only send you the papers after I recover from my heart condition.

F. Ruffini to A. Einstein

Turin, 8 November 1931

Honored Sir and Friend,

Allow me to make a plea on behalf of myself and very many, if not all, of my Italian colleagues.

The Italian Government has passed a law requiring all university professors to take a new oath, in which they pledge themselves to be <u>loyal to the Fascist regime</u> and, moreover, to exercise their profession with the intention of forming citizens devoted to the <u>Fascist regime</u>. It should be noted that such a serious commitment is imposed only on university professors, unlike lower school teachers and all other state employees.

Neither I nor my son, who is a professor of the history of law at the University of Perugia, intend to take such an oath. And the same refusal will be given to the government's request by some of the most esteemed professors in the various universities. But most of the professors will have to bow their heads, because their modest and often difficult economic situations do not allow them to face the sanction that will follow the refusal, that is, dismissal from their posts.

We are left with only one hope, and that is that, if ever a voice of solidarity and protest arises from the most distinguished teachers of foreign universities, the government will withdraw its ill-considered decision, or at least will not repress those who refuse to take this oath.

I have taken the liberty of addressing myself to you, whose authority is so highly recognized throughout the European scientific world: see for yourself if you can do something to help your colleagues in Italy; and accept as of now the expression of our deepest gratitude.

Francesco Ruffini

G. Ferrero to A. Einstein

Rue de l'Hôtel de Ville, 8
Geneva (Switzerland)
10.XI.31

Dear Master,

Senator Ruffini sent me the enclosed letter through a friend, asking me to send it to you. You will see what it is about.

Could you, as Mr. G. Murray has already done and as I believe Mr. Painlevé will do, write to Mr. Rocco expressing the painful impression this measure has made on the intellectual world and asking him to intervene with Mr. Mussolini?

If there is a case in which intellectual cooperation should intervene, it seems to me to be this one. I do not know what use Intellectual Cooperation can be if the part of Europe that still has intellectual freedom does not help the part that has lost it to regain it. It would be good, moreover, for the Italian government to know that membership of the International Committee for Intellectual Cooperation also implies duties.

Your intervention, like that of Mr. Painlevé and Mr. Murray, will be very useful. On this point I can assure you, and alas! I am familiar with the conditions there.

Please accept, Dear Master, the expression of my best regards.

Guglielmo Ferrero

A. Einstein to G. Ferrero

Berlin W.30, 16 November 1931
Professor G. Ferrero Geneva
Rue de l'Hôtel de Ville 8

Esteemed Colleague,

At your and Ruffini's suggestion I sent a letter, here enclosed in copy, to Signor Rocco. I beg you for the moment not to make public use of this letter. This may still happen if Mussolini should pursue his objective. I have avoided making this threat in my letter because I believe that I would thereby have done harm to the good cause. However, perhaps it would not do any harm if the powerful were to learn unofficially that the publication of such letters is looming if this rightful plea is not fulfilled.

   With best wishes for the success of your defensive strategy, I am yours

A. Einstein

A. Einstein to Minister A. Rocco

Berlin W.30, 16 November 1931
Haberlandstr. 5
To Mr.
Minister Rocco
Rome

Dear Colleague,

Two of the most distinguished and esteemed men of science in Italy have turned to me in distress and asked me to write to you in order to avoid, if possible in this way, a ruthless harshness that hangs over Italian scholars. It is a matter of preventing a form of oath in which one must swear allegiance to the Fascist system. My prayer is that you advise Mr. Mussolini to spare this humiliation to the flower of Italian intelligence.

   However different our political convictions may be, I know that we agree on *one* fundamental point: we both regard and love the achievements of the development of European thought as great benefits. These are based on freedom of thought and teaching, on the principle that the search for truth must take precedence over all other aspirations. Only on this basis could our culture arise in Greece and celebrate its rebirth in the Italy of the Renaissance. This higher good was paid for with the blood of pure and great martyrs, for whom Italy is still loved and admired today.

   Far be it from me to discuss with you what interventions in human liberty may be justified by reasons of State. The search for scientific truth, however, freed from daily practical interests, should be sacred to all State powers; and it is in the supreme interest of all that the loyal servants of truth be left in peace. This is also doubtless in the interest of the Italian State and of its prestige in the eyes of the world.

In the hope that my prayer will find your kindly understanding, I remain.
With kind regards

Yours
A.E.

G. Righetti to A. Einstein

Commissione Nazionale Italiana
per la cooperazione intellettuale
7288 C.N.I.
Rome, 12 December 1931-X
Professor Albert Einstein
Berlin, W.30
Haberlandstrasse 5

Distinguished Professor Einstein,

My professor, H.E. Alfredo Rocco, who went on 30 November to Paris for the executive committee of the International Commission for Intellectual Cooperation and who was also away from Rome afterwards, would very willingly have answered before his departure your kind letter of 16 November concerning the oath required of professors in state universities.

It was not possible for him to do so also because he would have liked to give you some concrete news since he acknowledges your great merit of having written with intense interest in science and intellectual cooperation. In order not to postpone his reply further, he instructed me to write to you as soon as I had the necessary information. For this reason I have the honor to communicate to you the following.

The new oath formula does not require professors to declare their political beliefs. They are asked only for obedience and respect for the current constitution of the Italian state, a constitution which, as a result of legislative reforms, has received a new order.

By demanding of the professors an oath of allegiance to the new regime, they are demanding of them an oath of allegiance to the laws of the constitution without thereby, I beg to emphasize strongly, *demanding that the professors adhere to this or that political group.*

That this is the interpretation to be given to the required oath formula was also explained at the time to the professors who had asked for clarification from the university presidents, and I am pleased to be able to tell you that of the 1200 or so full and pre-tenure professors, only 7 or 8 had any qualms about this formula.

All the others (and among them there are some who are not Fascists and who in private life are even anti-Fascists) took the oath without difficulty. So I can give you for example also the name of the famous mathematician Levi–Civita of Rome University.

Even professors at free universities, who were legally excused from it, were willing to take the oath.

Allow me to express the hope that I have succeeded in dispelling the concerns, regarding which you have interceded, in the light of the cordial cooperation which, under the guidance of Professor Rocco, animates the members of the International Commission for Intellectual Cooperation.

Please accept, most illustrious professor, my highest esteem

Your devoted Giuseppe Righetti
(Dr. Giuseppe Righetti, Member of Parliament)

A. Einstein to G. Righetti

(Einstein's handwritten draft of his reply at the bottom of the letter sent to him by Righetti on 12 December 1931).

I should like to thank you and Mr. Rocco for your reply to my letter, even though unfortunately it was unsuccessful. Incidentally, I cannot help but admire the skill with which you use a language which is foreign to you.

With the highest esteem

A.E.

P. Bonetti to A. Einstein

Scientia
La Segreteria Generale
Milan, 6/2/1932
Prof. Albert Einstein
Faculty of Philosophy
University

My Dear Professor,

We are very sorry not to have published for so long now an article due to your illustrious pen.

We hope, however, that in the future you will not refuse an article for "Scientia".

We would really like to host on our pages a debate, carried out by the greatest physicists and mathematicians, who are after all the greatest thinkers, on the question of determinism, through articles accessible to the vast cultured, but eclectic, public of our Review.

It is for this reason that I take the liberty of addressing you (as well as Messrs. Heisenberg, Bohr, Dirac, Perrin, Planck, Schrödinger and a few others) as one of the most distinguished intellectuals of our time.

Prof. Enriques (in whose name I am writing to you) would particularly like your collaboration; he sought you out in Berlin, on the occasion of his recent stay in that city, in December, because he wished to make this proposal to you personally. He was very disappointed to learn that you were in America at that time.

Can we hope for this much desired article?

It would be a short article of about ten printed pages (3400 words). By which date could you possibly send us the article? We very much hope to have your cooperation for this occasion!

While waiting for your reply, please accept, Sir and illustrious Master, the distinguished greetings of Prof. Enriques and the expression of my sentiments of the highest esteem.

Paolo Bonetti

A. Einstein to Reale Accademia Nazionale Dei Lincei

Institute for Advanced Studies
Princeton, New Jersey,
3 October 1938
Reale Accademia Nazionale dei Lincei
Rome

I learn from newspaper reports that the expulsion of Italian Jewish scholars from your academies has been ordered. I allow myself the kind request to know if this news corresponds to the truth.

With the highest esteem.

Professor Albert Einstein
Registered mail!

For the record by order of the President since the matter in question constitutes an intervention which does not directly concern the writer.

(Handwritten Note at the Foot of the Letter)

A. Einstein to Reale Accademia Nazionale Dei Lincei

Institute for Advanced Studies

Princeton, New Jersey,
15 December 1938
Reale Accademia Nazionale dei Lincei
Rome

Distinguished Gentlemen!

I hereby request that you remove my name from your list of corresponding members!
    With the highest esteem

Professor Albert Einstein

Reale Accademia Nazionale Dei Lincei to A. Einstein

Rome, 2 January 1939—XVIIth

My dear Professor,

I acknowledge your resignation as a Socio Straniero of this Royal Academy.

Respectfully

    The President.

A. Einstein to B. Croce

Princeton, 7 June 1944

I learn that a person from here, who had the good fortune to visit you, declined to leave you the letter that I addressed to him but which was written to you. However, I console myself with the thought that you are now occupied with incomparably more important occupations and sentiments, and particularly with the hope that your beautiful country will soon be freed from the evil oppressors from without and within. In this time of general upheaval may you be permitted to render your country an exceedingly valuable service, for you are one of the few who, standing above the parties, have the confidence of all.

If ancient Plato could in any way see what is now taking place, he would feel at home, for, after a long course of centuries, he would see what he had seldom seen, that his dream of a government ruled by philosophers is being fulfilled in a certain way; but he would also see, and this with greater pride than satisfaction, that his idea of the circle of forms of government is always in progress.

Philosophy and reason themselves are far, for a foreseeable time, from becoming the guides of men, and they will remain the most beautiful refuge of elect spirits; the only true aristocracy, which oppresses no one, and in no one stirs envy, and of which those who do not belong to it cannot even recognize its existence. In no other society are the bonds between the living and the dead so lively, and our fellows of former centuries are with us as friends, whose sayings never lose their attractiveness, their fruitfulness, and their personal magic. And, finally, those who really belong to that aristocracy, may indeed by other men be put to death, but not offended.

With respectful greetings and best wishes.

A. Einstein

B. Croce to A. Einstein

Sorrento, 28 July 1944

Distinguished Friend,

Your letter was very dear to me, because I have always kept in my memory the long conversation we had in Berlin in 1931, when we shared the same anxious feeling about the danger to freedom in Europe: a commonality of feeling and intentions that I saw confirmed when I found myself collaborating with you—made an exile from your homeland for the furious fight against freedom—in the volume of essays on freedom (*Freedom*), prepared, four years ago, in New York.

Of the two theories of Plato, which you recall, that perfect republic, constructed and governed by reason and philosophers, has not, in truth, been received, or rather has been rejected, by modern thought; but the other has been retained, which was not peculiar to him, of the circle of forms, that is, of the necessary forms in which history perpetually moves: with this in addition that the circle was enlightened by the complementary idea of the perpetual advancement and elevation of mankind through the necessary course, or, according to the image which pleased your Goethe, its "spiral course". This idea is the foundation of our faith in reason, life and reality.

As for philosophy, it is not strict philosophy if it does not know, through its office, its limit, which is to bring to the elevation of humanity the clarity of concepts, the light of truth. It is a mental action, which opens the way, but it does not presume to substitute itself for practical and moral action, which it can only solicit. In this second sphere, it is up to us, modest philosophers, to imitate another ancient philosopher, Socrates, who philosophized but fought as a Hoplite at Potidea, or Dante, who poetized but fought at Campaldino, and, since not everyone, and not always, can perform this extraordinary form of action, to participate in the daily, and harsher and more complex warfare, which is politics. I too frequent the company, of which you speak with such noble words, of those who have already lived on earth and left us their works of thought and poetry, and I am reassured and refreshed in them. From time to time I immerse myself in this spiritual bath, which is almost my religious practice. But in that bath one is not allowed to stay, and from it we must leave to embrace the humble and often ungrateful duties that await us at the door.

Therefore I feel myself today, in conformity with my convictions and my ideals, committed to the politics of my country; and I would like, alas, to possess for it in abundance the forces that are most directly necessary to it, but nevertheless I give it those, whatever they may be, that I succeed in gathering within me, even if with some difficulty. And I thank you for the generous good wishes that you make for Italy, which has suffered a sad and painful fate prepared by the confusion produced in it as in other countries by the previous war in which it was possible for foolish and violent people to seize the powers of the State, not without the great applause and admiration of the whole world, and to turn and force Italy onto a path that was not its own, that all its history belied. Because never has Italy, since the fall of the Roman Empire, aspired to dominion in the world, and for centuries it has implemented or sought freedom and has been unified in freedom, and its nationalism and fascism has come from foreign concepts, which only those foolish and violent people could adopt as a pretext for their mischief. Not even ancient Rome had such a delusion, because its work was to continue that which had been so brilliantly begun by Hellas and to create a Europe, giving civil laws to the barbarians who had none or who had barbaric laws.

War is war, and obeys no principle other than its own, and even the noblest ideologies are means of war, as every connoisseur of history knows and every wise man understands. When the war is over, the internal struggle for civilization and freedom will take place no less in the victorious countries than in the vanquished ones, all of which have been shaken by the war they have sustained and all of which are more or less unaccustomed to freedom. But since wars aim, as is their natural effect, at an arrangement of peace, it is to be hoped and recommended that the statesmen, who now direct them, should think as of now not to prepare such conditions in the various countries as would make a solid peace impossible, and, thus damaging the very cause of liberty, would prepare a new war, which can never be prevented by mere coercion, but requires the disposition of minds to peace, concord and the dignity of labor. "Tongues bind swords," as an old Italian philosopher used to say.

But I don't want to bore you with entering into a discussion of what I observe and judge in the affairs of international politics, with particular reference to Italy; for indeed I should also ask your pardon for having taken occasion from your kind and cordial words to expound my thoughts on the lofty questions that you touched on. But "naturam expelles furca, tamen usque recurret": that is, the nature of the philosopher, who distinguishes and theorizes. And, thanking you for your kind letter, I shake your hand.

Your
B. Croce

A. Einstein to G. Castelnuovo

To the President of the
Academia Nationale dei Lincei
Rome, Italy
June 26, 1946

Sir

With great pleasure I see from your letter of April 26th 1946 that the Academia Nationale *[sic]* dei Lincei has resumed its activities for the benefit of science, your country having been liberated from fascist oppression.

My mailing-address (private) is: 112 Mercer Str. Princeton N.J.

Faithfully yours,
Albert Einstein

P.S. Dear Dr. Castelnuovo:

I shall be happy indeed to become again a *socio straniero* of your Academy as I have been in the good times of the past.

Saluti affettuosi

A.E.

A. Einstein to E. Marangoni

16. VIII.46

Dear Ernestina!

I was happy to see a letter from you after so many years—and what years! I am also glad to hear, that all the Casteggiani friends except the husband of Julia Mai are unharmed—and dear Mussolini, ⚰ as honestly deserved. The happy months of my sojourn in Italy are the fondest memories. Your father in the middle like a second Leonardo da Vinci. Days and weeks without anxiety or tension. Many greetings to dear Mrs. Julia who has suffered and still suffers more than many others. I settled in America already in 1932 lacking true trust in men and in God—a wandering life with only one constant—mathematical work.

I saw with pride your confidence in my power over the material world—illusionary confidence. My relations with British officialdom are explicitly cold because I have publicly accused the British colonial regime on the occasion of the Palestine problem.

You will understand very well that under these conditions my blessing would not be very effective for the renaissance of the bridge over the Ticino at Pavia. And I would do so gladly, if I saw any possibility of success.

Best regards and best wishes to all.

your Albert Einstein.

(With desperate help from Maja who is bedridden with a broken leg in Princeton Hospice).

A. Einstein to Mrs. Lanzoni

Mrs. Livia Lanzoni
Via Orcagna 51
Florence, Italy
April 25, 1949

My dear Mrs. Lanzoni:

Thank you very much for your kind letter of March 23rd.

The important investigations of your father together with Levi–Civita have helped me considerably in my work concerning the general theory of relativity.

Yours very sincerely,
Albert Einstein

A. Einstein to Mrs. E. Marangoni

October 1st, 1952

Dear Marangoni:

I thank you for your two letters. The gentle reproach I feel expressed in your second one is not justified, because you did not give your address and your married name in your first letter. Therefore I wrote you to <u>Pavia</u> without street etc. and addressed to Marangoni. This letter came back, of course. I hope, however, that the post will find you in Casteggio.

What beautiful memory is Casteggio. What charm has the little town seen with the admiring eyes of youth. I see again before me your father who so benevolently looked upon the world and who seemed to resemble Lionardo *[sic]* da Vinci. I see gain *[sic]* before me also lovely Samazaro and Signorina Mai. It is difficult for me to realize that we all have grown old together. In the imagination the distance fixes everything as it was at that time.

I did not know that you have relations with the Besso family (Michele Besso is an old friend of mine) so that you know about the death of Maja and Pauli. Maja has lived with me the last 12 years of her life to avoid the Nazis. Her dream was to return to Italy but I knew that her illness would not permit this anymore. But she was cheerful and of vivid mind until the last and did always try to keep up the bonds with her old friends who all were touchingly attached to her. I, myself, on the other hand always loved solitude, a trait which tends to increase with age. It is a strange thing to be so widely known and yet be so lonely. But it is a fact that this kind of popularity—as it has become the case with me—is forcing its victim into a defensive position which leads to isolation…

We had to witness gigantic <u>political</u> upheavals and shall witness still more if we are not called away in time. Essentially everything is always the same. The nations always walk again into the trap, because the atavistic impulses are stronger than reason and aquired *[sic]* convictions. Old Lichtenberg said rightly: "Experience does not make clever, because each new folly appears in a new light".

With cordial regards and wishes,
yours,
Albert Einstein.

A. Enriques De Benedetti to O. Nathan[5]

Lungo Po Armando Diaz 8
Turin (Italy)
Turin (Italy), 17 February 1957

Dear Mr. Nathan,

Excuse me if I do not know your language and I have to write in Italian.

I am the daughter of a scientist (Prof. Federigo Enriques, mathematician and philosopher, who died in 1946). My father had many contacts with Albert Einstein; he was in our house in 1921 and I always remember the wonderful hours spent with the great Genius. He had come with one of his sons.

He gave three lectures at the Archiginnasio in Bologna plus a conversation to which he had invited only young people. Although I was then only 19 years old, he was very kind to me. Perhaps he had read such enthusiasm in my eyes. He wanted me also to go to the conversation reserved for young people. This conversation was very interesting; we spoke above all of philosophical problems, of space considered as unlimited, but not infinite.

He came to luncheon with us twice; he told us about his family and his passion for the violin. Walking with him through the streets of Bologna, we noticed that he also had a fine taste for art, that he knew how to distinguish the ugly from the beautiful. His eye rested on small ancient, hidden monuments, and he knew how to understand the style and appreciate the beauty of each one.

But the most interesting thing I have to say about Einstein is his relations with Italy, as they appear from a letter from him that I still keep and from my very clear and precise memories of what happened afterwards. When mankind had the misfortune of losing Einstein, I wanted to honor his memory by writing in an Italian periodical all that I knew about the dialogs between him and my father. My father was a professor in Bologna until 1922, after which he had a chair in Rome. It was precisely after 1922 that what I have described in my article happened.

I therefore send you the newspaper "L'Europeo" where you will find all the news concerning Einstein's relations with Italy. In this article you will also find a letter from Einstein translated into Italian. The original is written in German. If you want,

---

[5] By kind permission of Prof. Andrea De Benedetti, Turin.

I can have it photographed and send it to you, but I guarantee that I have translated it word for word without changing anything. The letter is addressed to my father.

I also have a small album of memories in which he wrote some words to me by hand, again in German; these words are also reproduced exactly in the article in the Europeo.

I can tell you again that Einstein had a very close friendship with Prof. Tullio Levi–Civita of Rome, who died in 1941. The widow of Prof. Levi–Civita must keep a lot of correspondence between the two scientists: She lives in Rome via Bruxelles 53 and her name is "LIBERA TREVISANI LEVI- - CIVITA"; the correspondence with Levi–Civita is, however, of a scientific nature.

I am at your disposal for further explanations and news and also to send you the photograph of Einstein's original letter. I would be happy if you would send me your book in return.

Please accept my regards.

Adriana Enriques De Benedetti
Lungo Po Armando Diaz 8
Turin (Italy)

Part III
The "Scientia" Investigation and Other Writings

Recent Electro-Magnetic Theories and Absolute Motion

Orso Mario Corbino

Browsing through the rich scientific production of 1906 in the field of experimental Physics one encounters several Memoirs which, because of their intimate connection with fundamental principles, deserve to be known by the readers of this Journal.

My aim is precisely to set forth, in a series of articles, the principal results of this research, pointing out their connection with the theoretical questions currently being debated. And, to begin with, I shall deal in this first issue with the masterful research by Kaufmann "on the constitution of the electron" (*Annalen der Physik*, 4ª series, T. 19, p. 487, 1906) of a very great importance not only for the question in itself, but for the consequences that derive from it with regard to the electrodynamics of bodies in motion and to the question of absolute motion.

By studying the conduction of electricity through liquids and gases, Physicists have been led to admit that electricity consists of definite elementary portions, which behave as *atoms of electricity*, and with respect to which all charges are multiples according to integer numbers, just as, in Chemistry, the mass of any quantity of a simple body is a multiple according to an integer of the mass of the relative atom.

The atom of electricity, which was intuited by Helmholtz and then taken by Lorentz as the basis of his theory, was more fortunate than the chemical atom since J.J. Thomson, who gave it the name of *corpuscle,* was able to determine its absolute electric charge by counting the number of free corpuscles existing in a given volume of gas, subjected to certain treatments.

The corpuscles, i.e., the *electrons*, as they are now called, are emitted in great numbers and with enormous velocity from the cathode in vacuum tubes, constituting cathodic rays, and from radium, constituting β rays; and since these rays result from electric charges in very rapid motion they must undergo, as experiment had already revealed, a deviation in an electrostatic field or in the presence of a magnet.

O. M. Corbino, *Le recenti teorie elettro-magnetiche e il moto assoluto*, "Rivista di scienza", 1, 1907, pp. 160–167.

On the other hand it had been deduced long ago that an electric charge in rapid motion must present a sort of reaction against the disturbing forces of motion, as a material mass does by virtue of inertia. The electron must therefore present an apparent mass, of electromagnetic nature, varying with the change of velocity; and precisely those changes must become perceptible to experiment when the velocity of the electron is great enough to be comparable with the velocity of light.

Max Abraham, of Göttingen, had calculated in 1902 the value of the apparent mass of the electron for different velocities, basing himself on the hypothesis that it is all of electromagnetic origin and that the electron in its movement, however rapid, preserves a spherical shape and an unchanged volume; this is the *so-called theory of the rigid electron.*

On the other hand Kaufmann had already, with some preliminary experiments (1900–1903), determined the values of the mass for different velocities by measuring the deviation of the radium β rays on which an electrostatic field and a magnetic field of known intensity act. It was precisely due to the very remarkable agreement of the results of the experiments with the values calculated by Abraham that it was possible to conclude that all the mass of the electron is of electromagnetic nature, and that therefore the electric charge of the electron is not associated with a material core.

It is to the problem of the mass of the electron in motion that all the electrodynamics of bodies in motion is reconnected and hence the mighty body of theoretical research by Prof. Lorentz, aimed, among other things, to seek the influence of the absolute motion of the Earth on the optical and electromagnetic phenomena observable on its surface.

All the electromagnetic theories that now compete in this field are founded on the hypothesis of an aether whose parts are immobile, one with respect to the other, and which is therefore considered as at absolute rest. In particular, Lorentz's theory admits that there are no electric charges other than those due to their movement; this movement would take place almost freely in conducting bodies, while it would be hindered by electrostatic forces in insulating ones.

And since the complex of electrons existing in a body at relative rest with regard to the Earth is pulled along by the latter in its motion through space, and is therefore equivalent to a system of convection currents, it seems at first sight that it must be possible, by the observation of the electric phenomena (or optical phenomena which are related to them) taking place at the Earth's surface, to deduce the absolute motion of the Earth, that is, its motion with respect to the aether.

However, all the experiments carried out to date to this end have always led to negative results.

Lorentz's first theory accounted for the negative outcome of these experiments as long as one looked for effects due to terms containing to the first degree the ratio of the speed of translation to the speed of light. It did not, on the other hand, justify the negative outcome of a famous experiment by Michelson and Morley which, because of the expected perceptibility of effects due to terms of the second order, should have given a positive outcome.

A new hypothesis was therefore necessary to bring the theory into agreement with this new fact; and since Poincaré had objected that new hypotheses would still have to be introduced to explain every new experiment with a negative outcome, Lorentz succeeded in his Memoir of 1904 in demonstrating, with some fundamental hypotheses and without neglecting terms of any order, that it is not possible to bring into evidence at the Earth's surface the motion of the Earth with respect to the aether considered as at absolute rest. To arrive at this result he had to admit:

1. that all bodies, including the electrons from which they result, animated by a translation velocity, contract in the direction of the motion, the other dimensions remaining unchanged. This contraction is a function determined by the ratio between the absolute velocity of the body and the velocity of light;
2. that the forces between non-electrified molecules, or between these molecules and electrons are modified for translation equally like the electrical forces in an electrostatic system;
3. that the mass of the electron is all of electromagnetic nature, and that the mechanical masses increase in the same way as the mass of the electron by virtue of motion.

With these assumptions we deduce a complete independence of all observable phenomena from the absolute velocity of the bodies in which they are located.

Moreover, from the hypothesis that the electron, spherical in conditions of rest, contracts in the direction of motion, the other dimensions remaining constant, a law of dependence is deduced between its mass and velocity different from that established by Abraham with the hypothesis of the rigid electron.

Lorentz succeeded, however, in proving that the numbers obtained experimentally by Kaufmann fit those calculated with his theory almost as well as those calculated with Abraham's theory.

Lorentz's premises and the consequences are reciprocal. In fact Einstein came to demonstrate that if the principle of relativity, i.e., the principle of the impossibility of ascertaining absolute motion, is placed at the basis of an electrodynamic theory of bodies in motion, then the Lorentz deformation must take place, with which the same law of dependence between the mass of the electron and its velocity is found. The same conclusion, of great importance for what follows, was reached almost simultaneously by Poincaré.

In the meantime it follows that Abraham's theory, which does not admit that deformation, is irreconcilable with the hypothesis that in no way is it possible, even by an experiment of concept, to reveal the effects of a uniform velocity of translation, i.e., absolute motion. This constitutes, according to some, a drawback of that theory; but on the other hand the hypothesis of deformation by motion forces one to admit either an internal potential energy in the electron, or a constant external pressure acting on the deformable and compressible electron; both of unknown nature and not ascribable to electromagnetic phenomena. So that it would be impossible to found on pure electromagnetic bases the mechanics of the electron and therefore the whole of mechanics. This is instead possible with Abraham's theory, and this possibility constitutes a merit of very great importance from the cognitive theoretical point of view.

Put in these terms, it is clear that accepting or rejecting either theory is tantamount to accepting or rejecting either of these claims:

1. It is impossible to found all of Mechanics on electromagnetic bases (Lorentz-Einstein-Poincaré);
2. It is not impossible to reveal absolute motion (Abraham); and the acceptance of the one implies the rejection of the other as long as one accepts the fundamental hypothesis, common to all theories, of an aether whose parts are at relative rest and which is therefore considered—by definition—as being at absolute rest.

Kaufmann's early measurements, which were not precise enough, did not allow one to decide between the two theories, since the divergences between theory and experiment (concerning the law of dependence between the mass and the velocity of the electron) were within the limits of experimental error.

It was for this reason that Kaufmann undertook new experiments designed to measure the deviations of a ray of radius β having different velocities, through the action of an electrostatic field and a magnetic field. This is not the proper place to report the details of Kaufmann's experiments, which constitute a true monument of refinement and precision. As regards the difficult and laborious calculations required to pass from photographs of the deviation curves to numerical values comparable with the various theoretical predictions, they were developed by a procedure quite different from that of Max Planck. The results that Planck communicated to the Stuttgart Congress last October are absolutely in agreement with those already obtained by Kaufmann.

By comparing the different theories Kaufmann came to the very important conclusion that Abraham's theory is confirmed, while Lorentz's is not.

In truth, as Planck has pointed out, even with Abraham's theory there are slight numerical differences which he does not believe to be entirely attributable to experimental errors. The discussion that followed the Stuttgart Congress after Planck's communication is very important. Planck, Kaufmann, Bucherer, Runge, Abraham, Gans, Sommerfeld took part in it.

Planck observed very appropriately that after all Abraham's theory and Lorentz's theory are founded on one or other of the two irreconcilable postulates: that of the exclusively electromagnetic concept of mechanics and that of the impossibility of ascertaining absolute motion. It seems, however, that he, out of a sentimental predilection, so to speak, for the Lorentz postulate, has been a little too pessimistic in judging that Kaufmann's experiments are not entirely decisive in favor of Abraham's theory on the grounds that even with this theory there are very small divergences, to which others have attributed much less weight.

For my part I observe that it would be opportune to abandon all metaphysical preconceptions and to judge the value of the postulates according to their agreement with the facts; nor can I share Planck's preferences for the principle of relativity, since even if Abraham's theory is not absolutely proved true by Kaufmann's experiments, certainly these are irreconcilable with Lorentz's theory and therefore with the postulate of relativity, as long as one admits an aether at absolute rest. That if one wants to use metaphysics, it seems to me that once one admits the aether at absolute rest, and bodies animated by a motion of translation with respect to the aether, it is in no way extraordinary that this motion can be revealed in some way more or less accessible to experiment.

On the other hand, we have already pointed out that the contrast between the theories under discussion is always subordinated to the fundamental hypothesis of an aether at absolute rest, without which the principle of relativity and a law of variation of the electromagnetic mass with velocity (i.e., a law of deviation of the β rays of radium) different from that predicted by Lorentz could exist together.

If this fact were to occur, that is, if the principle of relativity and a law of variation of the mass not corresponding to the Lorentz formulae were both proved, one would be induced to renounce the concept of an absolute aether, which seems absurd to Prof. Enriques (See: "Problemi della Scienza", Chap. VI), but for its simplicity, remains at any rate today the only conceivable basis for a concretely developed electro-magnetic theory.

Returning to the constitution of the electron and its shape during motion, while Kaufmann's experiments exclude, as we have seen, Lorentz's particular deformation, they do not demonstrate definitively that the electron remains of invariable shape and volume during motion. In fact the results of experiments also agree well with the predictions of another theory, that of Bucherer-Langevin, which supposes instead a deformation of the electron consisting in a contraction in the direction of motion and in a dilation in the transverse direction, with the original volume remaining constant. This theory also has in common with Abraham's theory that it permits an electro-magnetic explanation of mechanics; and although it is irreconcilable with the principle of relativity, since it admits in the deformed electron a ratio of the axes equal to that which occurs in the Lorentz deformation, it is not improbable that the revelation of absolute motion is thereby excluded, at least up to the limit of the experimental attempts performed so far. If so, ascertaining absolute motion would be experimentally, if not theoretically, impossible, and any contradiction would thus be avoided.

The deformations of Lorentz and Bucherer are not the only ones that have been proposed; in this regard it is worth reporting an interesting investigation by Prof. Righi carried out with the brilliant lucidity of thought that characterizes this illustrious Italian physicist.

In the meantime Prof. Righi notes that it would be difficult to conceive and calculate a distribution of charges at the surface or in the volume of the electron, when for us the total charge of the electron has become the final *indivisible* unit of quantity of electricity.

He observes besides that in deducting the apparent mass of the electron by equating its living force with the total electrostatic and electromagnetic energy of the field produced by it, it is necessary first to subtract from this energy the part that it possesses, as electrostatic energy, when it is at rest.

The boldest novelty in Righi's treatment concerns the concept of the volume of the electron.

Let us consider the electron as an electrified geometric point, instead of as a solid of definite form and volume. Then the electric force produced by the electron at a point in space grows, tending to infinity, as the distance of the point from the electron decreases to zero, and the same can be said of Maxwell's tension along the line of force.

Now let us admit that this tension, from a new property of the aether, cannot exceed a certain limit, so that if the tension at a point exceeds that value, the aether remains profoundly modified in its properties like a wire subjected to a weight greater than its breaking load, i.e. the tension therein cancels out, or, more generally, that the tension retains a finite value limit φ.

Then the aether may be considered in the usual way outside a closed surface surrounding the electron, and which is the locus of the points at which the calculated force has the value φ, while on the inside of that surface the form will everywhere acquire the value zero or retain the constant limit value φ.

It will then be possible to consider the shape and volume of the portion of the aether modified in its properties, that is to say, of that portion which is limited by the above surface, as the shape and volume of the electron.

This is a spherical surface for very small velocities of the electron, and in general it is a surface of the sixth order having as its axis of revolution the direction of motion, and presenting a polar diameter contracted with respect to that of the sphere, and a slightly dilated equatorial diameter. On these bases the calculation of the apparent mass as a function of velocity is developed. The results refer only to the longitudinal mass and to a rectilinear and uniform motion.

The value of the contraction of the electron polar axis turns out to be identical to that admitted in the Lorentz theory, but the overall deformation is different, as it is also different from that of Bucherer.

It would certainly be very interesting to push the results of the new theory further to enable it to be compared with experiment, and to establish its position in the face of the two postulates now contending for the field.

But the notion introduced by Righi of a force-limiting surface instead of the external surface of the rigid or deformable electron of the other theories constitutes evident progress. Therewith disappears the notion of electric charge distributed and rigidly connected in a portion of space that is not matter, is not aether, and is not even electricity if the distribution is superficial; but that had the only purpose, at least originally, of avoiding in the calculation that the force became infinite at some points.

The Principle of Relativity and Optical Phenomena

Guido Castelnuovo

Introduction

When Newton, in his immortal *Principia,* posited the notions of absolute stillness and absolute motion, he was obeying the spirit of the times rather than the requirements of a logical development of mechanics. The motion of a body is determined only when we assign a second body, or *system,* to which that motion can refer. On the other hand, the system can change with a certain arbitrariness, without altering the laws of motion. These only state properties of relative motions, and never allow absolute motion to be characterized.

The conclusion about the relativity of all observable motion, to which the construction of mechanics has led, is in full agreement with the results of modern philosophical criticism. It may be convenient to speak of a body at rest, but the expression, understood in Newton's sense, has no real content. Although these ideas are now familiar to the scientific world, yet so powerful is the tradition, so strong still is the need in many to find a foothold in the eternal motion of the universe, that the ancient metaphysical concept sometimes resurfaces and often informs the language of astronomers and physicists. One concession is truly made to the modern spirit. Next to the material bodies, which enter into the question under consideration, a hypothetical medium is introduced, the aether of optics and electromagnetism, with respect to which the motion of matter is defined. But then such properties are attributed to the aether so as to reconstruct in it that system in absolute stillness from which it was believed one had been freed.

G. Castelnuovo, *Il principio della relatività e i fenomeni ottici*, "Scientia", 9, 1911, pp. 64–86.

The history of optics is full of similar more or less veiled attempts. The renewers of wave theory, Fresnel and Arago, investigate the reason why light phenomena are not disturbed by the motion of the Earth. It is not really the problem of the relativity of the phenomena that concerns them. The construction of the great edifice was too laborious for them to have time to philosophize! It was a question of determining what properties should be attributed to the aether, in order that the theory might agree with the facts. Is the aether motionless in space, as Fresnel believes? Then why does no phenomenon reveal the aether wind that affects the Earth in its annual movement? Or is the aether transported by the Earth, as Stokes thinks? And if so, how do we explain the peculiarities of the aberration of light?

The great controversy, which, thanks to Maxwell's powerful synthesis, also involved electromagnetic phenomena, dominated most of the last century. It was down to the genius of Lorentz to establish a perfect agreement between the electromagnetic theory of bodies in motion and the experimental facts. For Lorentz the aether is at absolute rest, but by virtue of ingenious compensations, the relativity of the phenomena is perfectly respected. One would say (if the joke were permissible in such serious matters) that an evil demon, after having created the admirable fabric of the aether, had tried by subtle devices to hide its structure forever from the mortal eye.

Lorentz's theory, perfect from the mathematical aspect, cannot however entirely satisfy a philosophical spirit. If no experiment is capable of revealing the movement of matter with respect to the aether, it must be possible to create a coherent theory of the facts, where the aether does not intervene. The aether, in Lorentz's treatment, plays a role analogous to that of reinforcement in the construction of a building. Further progress should lead to the description of the building without reference to the reinforcement.

Progress was made by Einstein; in his wonderful works on the subject the *absolute* has disappeared, and relativity triumphs. The triumph is not achieved, however, without grave sacrifice. Our ideas about space, about time, about the simultaneity of facts are profoundly disturbed, to such an extent that many feel perplexed in accepting the new views.

Whatever fate, however, may befall them in the future, so great is the interest of the subject, that every student of natural philosophy may wish to know it, even if only in a superficial form. To this end this paper is devoted. In a first part, of inductive character, it is shown how the impartial examination of the elementary facts of optics leads to the formulation of the fundamental principles of Einstein's treatment; and in a second part it is shown how those principles, following a deductive path, lead to a reconstruction of the kinematics different from the one that Galileo and Newton handed down to us.

The Principle of Relativity

When studying a phenomenon happening in any body, it is often convenient to connect the body to a reference system. The choice of the system is not entirely indifferent. Experiment reveals to us the existence of systems with respect to which the laws of mechanics assume a particularly simple character. Such a system is provided for instance by the Sun, if we abstract from the motion of rotation of the body around its axis. Every other body, or system, that is animated by a motion of uniform translation with respect to that one possesses the same requirements. Each of these *natural systems* of reference has in dynamics the same role as the space at absolute stillness of the metaphysicians; no mechanical fact authorizes one to prefer one of the said systems to the others.

Let us now suppose that in a body endowed with a uniform motion of translation (in the above sense) a phenomenon takes place over which no influence of external bodies is exerted. Experience teaches us that the peculiarities of the phenomenon are quite independent of the motion of translation of the body and can give us no clue as to the motion itself. Thus an experiment on the Earth, whose duration is short enough for us to abstract from the motion of rotation and the curvature of the orbit described by our planet, tells us nothing about the direction and velocity of the path we describe around the sun. But if the phenomenon considered on the Earth depends on the presence of other bodies, as, for example, the tide depends on the attraction of the Moon and the Sun, we will be able to obtain data about the relative motion of the three celestial bodies, not about the motion of the solar system in space.

In short: *no phenomenon occurring between several bodies, whose motions with respect to a natural reference system are known, allows one to decide whether the system itself is at rest or at motion.*

This is the *principle of relativity*, which all mechanical phenomena satisfy. There is reason to believe that its field of application is much broader, that it is really a natural law valid for every order of phenomena, as the statement itself declares. But since the consequences of that law, in certain regions of physics, seem to conflict with the results of classical theories, the verification of the principle of relativity, appropriately specified, is not superfluous.

The question arises for optics (and for electrodynamics, of which this is a particular case). It is to establish whether an optical phenomenon, in which we take into account *exclusively* the *material bodies* that take part in it and the *material media* through which the light passes, can give some clue about the motion of those bodies or of those media with respect to a *particular* reference system (the aether at absolute rest). Only when the answer is negative, we will say, with Einstein, that the phenomenon satisfies the principle of relativity.

Optical Phenomena Within a System at Relative Rest

Let us suppose, for the sake of clarity, that the source of light, the mirrors, the prisms on which the light is reflected or refracted, are fixed on the surface of the Earth. Do the particularities of the phenomenon depend on the angle that the light ray forms with the direction of the motion of the Earth along its orbit?

Arago's ingenious spirit proposed the question, and attempted to answer it with the means available to physics at the beginning of the last century. The answer was negative; he communicated it to Fresnel, inviting him to find the explanation on the basis of the wave theory then being formed. Fresnel, in a famous letter of 1818, observes that the most natural explanation would be obtained by supposing that the aether was transported by the Earth in its motion; but he does not stop there, since that hypothesis seems to him incompatible with the phenomenon of astronomical aberration. Accepting the opposite hypothesis, Fresnel is led to ask what law should regulate the propagation of light in moving material media, so that the refractive indices of bodies are not affected by the motion of the Earth, as required by Arago's experiments. The research has led to the discovery of the drag coefficient of waves in material media, which we shall have to discuss later. We wish for the present to remain in the domain of facts. We will therefore limit ourselves to observe that the experiments of Arago and of the physicists who with greater experimental means followed the same direction (among others Mascart) led to the following conclusion. The laws of reflection, refraction, polarization… of light are not affected in the least by the motion of the Earth.[1] Geometric optics, on the Earth's surface, entirely satisfies the principle of relativity.

However, one important question remains to be resolved. Does the velocity of light emitted by an earthly source, and measured with respect to the Earth, have a constant value c (about 300000 km/s) in all directions, as the principle of relativity requires, and as would be the case if the aether were pulled along by the Earth? Or does the said velocity compound with the velocity v (about 30 km/s) of the Earth along the ecliptic, and does it vary therefore between the two limits $c - v$ and $c + v$ according to the direction of the ray, as the hypothesis of the aether at absolute rest leads to? The experimental results reported above say nothing about this, since they agree with both the one and the other hypothesis. Therefore a new experiment was needed.

The experiment, admirable for its simplicity of concept and accuracy of technique, was proposed and performed by Michelson in 1881 and repeated six years later by the same physicist with Morley. A horizontal ray of light, on meeting a vertical plate of glass at an angle of $45°$, splits into two, one of which runs along the prolongation of the first, the other in a perpendicular direction. These two rays, by means of appropriate reflections, are led back to their paths and finally forced to overlap. The

---

[1] A single experiment, performed by Fizeau, on the plane of polarization of a ray of light forced to pass through a series of glass plates appeared in contrast with the preceding statement. But the result seemed doubtful to Fizeau himself, and to Maxwell and Rayleigh who discussed the experiment, which has not so far been repeated.

phenomena of interference to which the two superimposed rays give rise make it possible to evaluate with supreme precision the difference in the times taken by the two rays to travel their respective paths, which, in the parts where they diverge, may be considered as the two catheti of an isosceles right triangle. The triangle can then be rotated around its vertex in such a way as to attribute arbitrary orientations to its catheti, one of which can be brought to coincide with the direction of the Earth's annual motion. The experiment was carried out with such accuracy that it would have made it possible to reveal whether two rays of light propagating in perpendicular directions had a difference in velocity of the order of a ten-thousandth of the total value (30 km out of 300000 km), such as was predicted by the Fresnel theory and the first form of the Lorentz theory. Well, the result was negative. The two velocities turned out to be equal, within the degree of approximation compatible with the means employed.[2]

It is therefore legitimate to conclude that on the terrestrial surface, light propagation occurs with the same speed in all directions, in conformity with the requirements of the principle of relativity.

The preceding conclusions are valid, as we have said, on the surface of the Earth, and, indeed, for light sources that are motionless with respect to the Earth. The philosophic spirit, however, has not been content with reaping the fruits of these memorable experiments. With that audacity which may lead to error, but which has often led to the greatest triumphs of science, it has sought to extend the results to the entire universe, asserting that optical phenomena everywhere satisfy the principle of relativity. It is admitted, in particular, that the propagation of light in vacuum takes place in all directions with the same velocity, for an observer at rest with respect to the source, whatever the state of uniform rectilinear motion of the source.

Only the progress of our astronomical knowledge will be able to tell us whether this audacity is justified, as various considerations at present lead us to believe.

Optical Phenomena Between Moving Bodies

In the above experiments we assumed that the light source, the observer and the material media through which the light passes were at relative rest. New problems arise in the opposite hypothesis.

Let us specify the question by imagining a schematic experiment. A ray emitted by a source S passes through an empty tube AB; two observers, located at the ends of the tube, measure the time taken by the light to pass from A to B, and deduce the speed of light with respect to them. How does this speed vary when the source and

---

[2] Taking into account the double path of the light beams along the two catheti, we see that, in the hypothesis of Fresnel-Lorentz, the difference in the times obtained by the experiment should have had, with respect to each of these times, a value of the *second order,* that is comparable with the square of the ratio $v : c = 30 : 300,00$. Precisely, attributing to the catheti the length of 11 m., we should have found a difference of 4.10^{-19} s. The possibility of measuring such an extremely small interval can give an idea of the degree of perfection to which optics has arrived!

the tube move independently along the line SAB? Classical optics would answer as follows: the speed does not vary if only the source moves, it has a certain value c if the observers A and B are motionless with respect to the aether, and has the value $c-v$ if they move with the speed v in the direction of the propagation of light. Now this answer is formally contradicted by the experience of Michelson and Morley, if the aether is assumed to be at absolute rest, and it is certainly incompatible with the conclusions stated above that have been drawn from that experiment. We are therefore led to believe that the required velocity cannot depend other than on the *relative* motion of the observers with respect to the source, in accordance with the principle of relativity. The law of dependence is unknown; the simplest hypotheses that may be made in this connection are these:

(a)   either the speed of light, within the tube AB, is a universal constant c, independent of the state of stillness or motion of the tube with respect to the source (Einstein's principle);

(b)   or the said velocity, equal to c if the tube is at rest with respect to the source, assumes the value $c-v$ when the tube moves away from it with the velocity v (ballistic hypothesis).

According to the first hypothesis an observer, who measured the velocity of the light rays coming from various sources in various conditions of motion, should always obtain the same value. Instead by virtue of the second hypothesis the value should be greater or lesser, depending on whether the source was approaching or moving away from the observer, as would happen for a bullet launched from a cannon in motion with respect to the target.

Between these hypotheses only experiment can decide. Unfortunately, it does not seem easy to give an experimental answer as long as one remains on the surface of the Earth. It was therefore thought to resort to extraterrestrial sources.

One way is suggested by the phenomenon of the aberration of light. As is well known, and as we shall later say, the direction in which a star is seen does not coincide with the straight line joining the star to the observer but forms a certain angle with it. The maximum value of 20″ 5 that this angle reaches in the course of a year (*aberration constant*) allows us to evaluate the speed of light coming from the star, since the sine of that angle is the ratio of the speed of the earth along the ecliptic to the speed of light. So if this could depend on the proper motion of the observed star, different values for the aberration constant would have to be found for different stars. For this purpose the celebrated astronomer W. Struve examined a series of observations made at Dorpat between 1818 and 1826 on the polar star and on a presumed companion of it, and believed he found a difference between the aberrations of these two stars due to the above cause. But the examination of more recent observations made by Nyrén in 1888 cast doubt on Struve's argument. On the other hand, even if we admit the ballistic hypothesis, a star rushing towards the Earth with the (already relevant) speed of 60 km/s would result in a decrease of only one five-thousandth in the value of the aberration. Now what astronomer can guarantee, in the measurement of an angle, the accuracy of four thousandths of a second?

A means of resolving the question could be provided by a systematic study of those double stars, which describe orbits in planes passing approximately through the Sun or the Earth (which is equivalent given the immense distance of the star). In the orbit of such a star there are two positions in which the star approaches or recedes from the Sun with the greatest speed. The instants in which the star passes alternately through these positions can be considered at first approximation as equidistant. But we observe, by means of the spectroscope, these instants delayed because of the time that the light takes to reach us from the star. Now in Einstein's hypothesis the delay would be constant, so that the *observed* instant of maximum movement away should still be equidistant from the *observed* instants of maximum approach that precede and follow it. In the ballistic hypothesis instead the delay would be smaller or greater, according to whether the star approached or moved away from the Sun, and therefore the time interval elapsed between the first and the second observation should exceed the interval between the second and the third. Indeed, the difference would have a significant value (several hours, and even a few days) as a result of the enormous distance of the star, as an easy calculation shows. The imperfect knowledge we have about the motions of double stars would not, to tell the truth, allow us to eliminate other possible anomalies. But if for all double stars under the above conditions a constant cause of irregularity appeared, there would be a strong presumption for attributing that cause to variations in the velocity of light.

For these problems the future will be able to give an answer. In the present state of science no fact authorizes one to prefer the ballistic hypothesis to the hypothesis that Einstein drew from Lorentz's electromagnetic theory.[3]

And since indeed the consequences of this seem in accordance with experiment, we are led to adopt the principle of the constancy of the speed of light with respect to any natural system of reference.

But we must immediately point out a singular consequence of this principle. It is in full contrast with the law of composition of motions enunciated by Galileo and adopted by classical mechanics. Einstein, by adopting the principle of the constancy of the speed of light, was in fact led to construct a new kinematics, in which the velocities are composed according to a different and more complicated law than the one ordinarily used. But we will have to return to this later.

---

[3] In the course of the publication of this paper, I read in the "Physical Review" two articles that concern the question dealt with here. In the first (No. of February, 1910) a communication made by Mr. Comstock to the meeting held by the American Physical Society at Princeton in October, 1909 is briefly reported. Comstock proposes the method of double stars, which I myself have suggested above, and adds that the research he has done so far by this route do not seem to agree with the ballistic hypothesis. In the second article (No. of July, 1910) Mr. Tolman expounds a simple observation to discriminate the two hypotheses. He, after noting that the Doppler effect presents different peculiarities in the two theories, deduces from the spectroscopic examination of the celestial and terrestrial motions that only Einstein's hypothesis is acceptable.

The Aberration of Light

This phenomenon, whose ingenious discovery is due to Bradley (1728), has had a relevant part in the discussion of the various hypotheses about the motion of the Earth with respect to the aether. But, in the order of ideas in which we have placed ourselves, the theory of aberration is of interest only in so far as it confirms the principle of relativity. On the contrary, we will show how, by availing ourselves of this principle, we arrive at the simplest explanation of that phenomenon.

A light source S (star) sends a ray towards the observer T (Earth), which for the moment we will suppose to move by uniform translation with respect to S. If the propagation of light were instantaneous, the star S would be seen by T in the real direction TS. In order to examine how the phenomenon changes in the opposite hypothesis, we can use indifferently a reference system connected with S or with T. Let us adopt the second choice, that is, let us consider the observer T as fixed and the celestial body S as mobile. A light ray leaves the celestial body at the instant in which it occupies the position S_0, and traveling along the straight line S_0 T it reaches the observer after a certain time t. In the meantime the star has passed from the position S_0 to the position S_1. The observer sees the star in the direction TS_0 described by the ray; but the real direction of the star at the instant of the observation is TS_1. The two directions form the *aberration angle*. It is the angle of a triangle of which the opposite side $S_0 S_1$ and the adjacent side TS_0 stand in the ratio $v : c$, of the velocity of S with respect to T (or, which is the same, of the Earth with respect to the star) to the velocity of light.

If the relative motion of S and T were translatory, the phenomenon we are talking about would escape the attention of astronomers, because only the apparent and not the actual direction is revealed by the telescope. Actually the Earth describes an orbit almost circular around the sun, the ecliptic. This causes the aberration angle to vary in magnitude and orientation in the course of a year, as if the apparent position of the celestial body were describing, around the real position, a circle in a plane parallel to the ecliptic plane, with radius vt (a circle that is projected on the celestial firmament in the form of an ellipse). The maximum value of the *aberration* angle is, as I already said, the *aberration constant* $\alpha = 20'' 5$, such that sen $\alpha = v : c$.

The theory of aberration, as expounded above, immediately accounts for a well-known paradox. Father Boscovich, relying on the latter formula, had expressed the idea that, observing the stars with a telescope filled with water, we should find for the aberration constant a value α' different from the previous one, such that sen $\alpha' = v : c'$, where $c' = 3/4\, c$ is the speed of light in water. The experience performed by Airy in Greenwich belied the prediction. And so it had to be, since the ray S_0T preserves its rectilinear direction even if it is forced to pass through a telescope full of water, whose upper and lower surface is normal to the ray itself; the apparent direction of the star does not depend therefore on the transparent medium contained in the telescope. Nor does the real direction TS_1 vary, since the time it takes light to travel the distance from the star to the observer is not appreciably increased by the effect of the water contained in the telescope. But if this could be given a dimension comparable to the said distance, the value of the angle α would be altered. Boscovich's prediction would only be verified when the telescope's lens could reach the star!

Units of Length and Time in Systems in Motion

The phenomena of which we have been speaking briefly give us some clues as to the laws which the optics of bodies in motion obeys. But a deeper knowledge of these can be obtained only by multiplying experiments and increasing their precision. On the other hand, in order to guide the experimenter, it is expedient to construct a theory, which from certain fundamental hypotheses deduces consequences that can be compared with the facts.

The theory alluded to is in essence due to Lorentz. But Einstein reconstructs it remaining in the order of the concepts developed up to now. The elementary part (kinematics) which alone we wish to discuss, rests on two principles. One is the principle of relativity, which we have already sufficiently clarified. The other relates to the constancy of the speed of light, and requires some further caution, since the evaluation of a speed implies the choice of a unit of measurement for space and time.

The choice, really, is indifferent as long as we refer to a certain system, as long as we operate, for example, on the Earth's surface. Einstein's postulate states then that the speed of propagation of light in vacuum, referred to that system, has a constant value, independent of the state of rest or motion of the light source and of the wavelength of the light under examination. This is a hypothesis that experiment will be able to confirm or deny. But when we are forced to compare two different reference systems, in motion the one with respect to the other, how will we be able to compare the respective units of measure? The transport of the units can be practically non executable, or give rise to alterations whose meaning cannot even be specified. It is convenient therefore, following a tendency familiar to physicists, to resort to natural units, and to choose for example as linear unit for each system the wavelength of a given radiation emitted by a source connected to the system, and for time unit the period corresponding to that wave. The propagation velocity is then equal to 1. Einstein's principle states that the same conclusion is valid for every light propagation in vacuum, whatever the reference system. The principle splits therefore into two parts: a hypothesis that can be submitted to experiment, and a pure convention depending on the choice of the units of measure.

Space and Time According to Minkowski

We will not follow Einstein in the construction of the new kinematics. But we would like to give an idea of the principal results, making use of a geometrical representation, which was put in a new light by the brilliant spirit of Herman Minkowski, whose premature loss science still mourns.

The movement of a point on a straight line x gives rise to a frequently used diagram, which is obtained by taking for the positions $x_1, x_2 \ldots$ occupied by the point at the instants $t_1, t_2, \ldots$, as many perpendiculars to the line x, of lengths $t_1, t_2, \ldots$ The extremes (x_1, t_1), $(x_2, t_2) \ldots$ of those perpendiculars form a line, which represents, so to speak, the history of the point through time. A fixed point will have as its image a line perpendicular to the *x-axis*; if in particular the point coincides with the *origin* O, from which the distances on the *x-axis are* counted, the line in question is the *time axis, t.* A point moving with uniform motion is represented by an oblique straight line, and so on.

It is clear, on the other hand, that such a procedure can be used to represent the motion of points or areas belonging to a plane $x\,y$, by means of lines or tubular surfaces of the space in three dimensions $x\,y\,t$; and it could be further extended to the description of the motions in ordinary space, if we overlook the difficulty of conceiving the figures in a space in four dimensions.

To this representation Minkowski ascribes a deeper meaning than that of a simple diagram. He imagines a spirit superior to ours, which conceives of time as the fourth dimension of space, and can follow the hero of a well-known Wells novel in his marvelous journey through the centuries. The movements, which we see happening above an $x\,y$ plane, appear to Minkowski's demon crystallized in the respective diagrams of $x\,y\,t$ space. He can embrace with a single glance the history of the universe, while we are forced to build it up painfully, climbing step by step an interminable staircase whose steps represent the hours.

This fantastic world that Minkowski evokes to our spirit suggests an image that we should keep in mind. Let us conceive in Minkowski's space $x\,y\,t$ a body extending to infinity, for example a permeable cylinder, a shadow-cylinder, as we shall say, borrowing a term suggested by cosmography.[4] Let the cylinder be oblique with respect to the horizontal plane $x\,y$ and leave a shadow trace there. Let us now imagine that the said plane moves parallel to itself as time varies, and pulls an observer along with it who describes the vertical axis t. Then the shadow section of the cylinder (which is fixed) will move with respect to the observer. If, however, the observer is not aware of the motion that pulls him along with the plane, he will attribute the visual perception to the actual motion of a disc on a fixed horizontal plane; to that motion, precisely, which has the cylinder as its image in Minkowski's representation.

Let there still be on the plane $x\,y$ a second observer, who, when the plane changes, describes, without noticing it, a new line t', parallel to those that generate the cylinder. This observer will consider the shadow immobile and the first observer in motion. In short, the two individuals will perceive the same phenomena differently, according to the well-known laws of relative motion in ordinary mechanics. The principle of relativity means therefore, in Minkowski's representation, that the axis of time can be chosen at will. In this context the straight line t' can be substituted indifferently for t, since of two observers it is indifferent to consider the first or the second at rest.

---

[4] In the figure shown, assume the sheet of paper $x\,t$ to be vertical, the *y-axis* (unmarked) perpendicular to it in O, and then the plane $x\,y$ horizontal; then assume that the cylinder cuts the plane $x\,t$ along the straight lines C$'$ C, D$'$ D.

Fig. 1

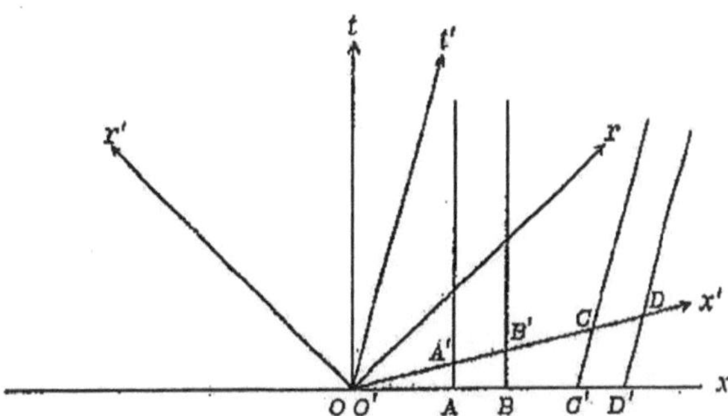

What is now the scope of Einstein's principle? For the sake of clarity, let us consider only the motion of a point on a straight line x, in which case the diagram is contained in the plane $x\,t$. (Fig. 1). The observer O at the instant $t = 0$ throws a light signal in both directions along the x-axis. If the units of space and time are chosen so that the speed of light is 1, the two propagations are represented in the diagram by the two bisectors r, r' of the angle $x\,t$. These lines represent the same phenomenon also for a second observer O', who moves with respect to O along the x-axis with a certain velocity v, and has as time axis a line t' different from t. Now, if the diagram that corresponds to it differs from the previous one only for the time axis, with the x-axis of space remaining unchanged, the propagation velocities of those light signals would be, for the new observer, $1 - v$ and $1 + v$, in agreement with ordinary kinematics. Asserting that the speed of light is always 1, whatever the observer, is equivalent to asserting that the change in the axis of time also brings a change in the axis of space, so that the straight lines r, r' are still bisectors of the angle $x't'$, as they were of the former angle $x\,t$.

Thus according to Minkowski the two different representations of the principle of relativity in Newton's kinematics and in Einstein's kinematics appear in evidence. In the first, the direction of the axis of time can vary at will, without changing the x axis, or the $x\,y$ plane, or the $x\,y\,z$ space, where we see the phenomena occur. In the second, on the other hand, the axis of time and space revealed to us by our senses are linked. Every change in the direction of that axis brings a change in the orientation of this space.[5]

The change would only be perceived by Minkowski's demon. But we should also notice some difference in the particularities of the phenomena if our senses were acute enough.

---

[5] The link between the axes x, y, z, t, and the new axes x', y', z', t' is such, that the transition formulas change the expression $x^2 + y^2 + z^2 - t^2$ into the analogous $x'^2 + y'^2 + z'^2 - t'^2$. From this condition we can deduce how the units of lengths and times vary as the axes change. E.g. on the figure shown, the units of measure exiting from O in the various directions would have the other extreme above one or the other of the hyperbolas $x^2 - t^2 = \pm 1$.

The Contraction of Bodies in Motion

A first observation is this: the same object presents itself in different forms, depending on whether it is examined by an observer at rest or in motion with respect to it. This is confirmed by the fact that the same figure of Minkowski's space, e.g., the shadow-cylinder discussed above, has different sections depending on the inclination of the secant planes (planes parallel to $x\,y$ or to $x'y'$, depending on whether it is the first or the second observer). Thus a *circular* disk or an immobile *sphere* with respect to a first observer gives to a second observer the impression of an *elliptical* disk or an *ellipsoid* in motion. A more detailed examination of the geometrical figure (taking into account the various units of measurement) would also provide the value of the deformation. While the segments perpendicular to the direction of motion keep their values unchanged, the segments parallel to the translation are contracted in the ratio of 1 to $\sqrt{1-v^2}$, v being the ratio between the speed of the motion and the speed of light. The contraction is really imperceptible for the motions that we see on the Earth or in the sky, which are slow compared to light propagations. For example, a sphere nearly as large as our Earth, and animated by the same speed of translation, would appear to an observer located on the sun as a flattened round ellipsoid, whose shortest axis would differ from the others by only 6 cm.

But contraction is relevant for bodies animated by velocities comparable with that of light (such as electrons). And it is such as to prevent a body from moving faster than light. In Einstein's kinematics there is therefore a limit that no speed can exceed. Or, in a more precise form, that kinematics does not apply to bodies animated by higher velocities, since for them the given definitions of unity of space and time become illusory.

The law of contraction of bodies in motion, already enunciated by Fitzgerald, was emphasized by Lorentz, who showed how it could serve to justify the result of Michelson and Morley's experiment in the hypothesis of the aether at absolute rest. But the same law has a quite different meaning in the theories of Lorentz and Einstein.

The former admits the aether at rest and imagines the units of length and time belonging to it. With these *absolute* units he measures the relative units belonging to a body in motion, and finds that the ratio between the latter and the former units is $\sqrt{1-v^2}$. A moving segment has therefore a length which depends on velocity. The maximum value is reached when the segment is at rest with respect to the aether.

The interpretation is quite different for Einstein. He does not admit absolute stillness, and can only compare bodies at rest or in relative motion. On a straight line let there be two segments AB, CD, the second of which moves with respect to the first with velocity v. An observer O, at rest with AB, measures this segment with his unit of length, and finds e.g. the value 1. Suppose that the same value is found for CD by a second observer O', at rest with this second segment. The two segments are then equal. But if the observer O tries, by means of optical signals, to measure the segment CD, which he sees passing in front of him, he already finds not the value 1, but the smaller value $\sqrt{1-v^2}$. And O' finds the same value if he tries to measure the segment AB.[6] The contradiction between the two judgments only means that the methods for comparing the two lengths are fallacious. The units belonging to two systems, in motion with respect to each other, are actually irreducible to each other. While the contraction for Lorentz is a fact, it is only a semblance for Einstein.

Local Time and the Concept of Contemporaneity

Similar observations occur when we compare the units of time of two systems in motion. The question is indeed the same for Minkowski's demon, but it assumes a different appearance for our earthly spirits.

It is easy to conceive how it is possible to tune two clocks that are at rest by means of optical signals exchanged in both directions. In the same way the clocks of a railroad network can be adjusted. The driver of a train, however, evaluates time according to another unit of measurement, the unit that belongs to the system in motion of which he is part. Now the two units are different. If, therefore, the various clocks had such exquisite sensitivity as to be able to evaluate trillionths of a second, the engineer would have to notice a discrepancy between his own time and that of the stations he passes. The moving clock lags behind the fixed clocks. If the time it takes the train to pass from one station to another is one hour (3600 s.) when measured on the fixed clocks, the time itself, estimated by the driver, is given by the smaller number $3600 \sqrt{1 - v^2}$, where v is the ratio between the speed of the train and the speed of light. The difference between the two numbers is extremely small for the speeds to which we are accustomed, and also for the motions of celestial bodies. Consider that an earth clock would lag behind a fixed clock on the sun by 15 s every century!

The observation is nevertheless remarkable, especially when we consider the physical meaning of the unit of time, which is the period of a given radiation. It turns out, in fact, that spectroscopic analysis should be able to reveal not only the motions that take place in the direction of the light ray, but also normal displacements, provided they are rapid enough. If our Earth travelled along its orbit with a speed a hundred times greater than it has, or if our instruments were ten thousand times more sensitive, a spectroscope pointed at the sun would allow this movement to be discovered. The Doppler principle thus assumes a more complex character in Einstein's kinematics than in ordinary treatment.

To every wandering star in the sky a certain *time* belongs, which the star can communicate to a body at rest with respect to it, to every point of the universe that is considered to be transported by it; this is the *local time* of that star. An observatory, situated on the Earth or on the Sun, could record over as many clocks the local times

---

[6] The figure shown should be interpreted as follows. When O's clock reads $t = 0$, he sees the first segment at position AB, and the second at position C′D′, and finds for that the value C′D′/AB. On the other hand, when the clock of O′ reads $t' = 0$, the second observer sees the first segment in the position A′B′, the second in CD, and finds for that the value A′B′/CD. Now, given the choice of length units on the *axes x, x′* we have C′D′/AB = A′B′/CD = $\sqrt{1 - v^2}$, where $v = tg\ xx'$.

of various stars. These clocks would all be in disagreement with each other, and all would be behind the clock that marks the local time of the observatory.

Now this disagreement brings a paradoxical consequence, which profoundly disturbs one of the most radical ideas of our spirit: the concept of the contemporaneity of phenomena. Let us suppose that a solar explosion and a magnetic perturbation on the Earth take place *simultaneously*; we mean (allowing for the time it takes for light to reach us) that the astronomer perceives the solar phenomenon 500 s after the magnetic needle has undergone an abrupt displacement. We are led to believe that any spectator lost in the universe, who could see the two phenomena, would still judge them to be contemporaneous, taking into account, of course, the delays due to the transmission of light. But this is not the case according to Einstein. Two clocks, one on the Earth, the other on the Sun, set to the local time of the spectator, who is supposed to be in motion with respect to the solar system, would not mark the same instant while the two phenomena take place. If the spectator travelled from the Sun to the Earth with the speed of 600 km/s, he would conclude that the solar explosion followed the magnetic perturbation by one second. He would arrive at the opposite conclusion if the direction of his motion were reversed.

In short, contemporary phenomena for one observer are no longer contemporary for a second observer who is in motion with respect to the first; contemporaneity is relative, like the values of lengths and times.[7]

The chronologies of the universe collected by the inhabitants of two celestial bodies wandering in the sky would present discrepancies which would not be trivial.

The Law of Velocity Composition and the Fresnel Drag Coefficient

It was already noted that Einstein's kinematics is incompatible with the law of composition of velocities of classical mechanics. If a bar AB passes above a straight line Ox and moves away with velocity u from the point O, and in the same direction a point moves above the bar with velocity v with respect to A, that point, according to Galileo and according to ordinary intuition, moves away from O with velocity $u + v$.

For Einstein the velocity of the point has instead the more complicated expression $(u + v): (1 + uv)$, if as unitary velocity we assume that of light.[8]

For the motions occurring on Earth the two formulas lead to the same result; for the motions of celestial bodies (some tens of kilometers per second) the results differ

---

[7] These and similar consequences result from the interpretation of the Minkowski diagram. For a first observer, whose reference system is xt (see figure), phenomena occurring simultaneously are represented by points on a line parallel to x. For a second observer (in motion with respect to the first), to which the axes $x'\ t'$ belong, simultaneous phenomena have as images points on a line parallel to x'. The geometrical fact that x and x' have different directions translates into the physical fact mentioned above.

[8] If we take for unity the velocity of 1 cm/s, the above expression becomes $(u + v) : (1 + uv/c^2)$, where c is worth 30 trillion.

at most by a few millimeters per second. But the disagreement of the two formulas appears, when one of the velocities approaches that of light. It already appears at once that the velocity of light in vacuum, compounded with any other velocity, gives as result the same velocity as the velocity of light, in agreement with Einstein's principle.

A very remarkable application of the new formula of composition of motions was made by Laue to justify the law of propagation of light in a liquid (or transparent medium) in motion. The speed of light in a liquid at rest is equal to $\frac{c}{n}$, where c is the speed in vacuum or air, and n is the refractive index of that liquid ($n = {}^4/_3$ for water). How does the result change if the liquid moves with known velocity v in the direction in which the light propagates?

Fresnel foresaw the answer by following a very ingenious theoretical path. To explain how Arago's experiments already mentioned do not allow the detection of the motion of the Earth, it is necessary to suppose that the speed of light in the liquid at rest increases, when the liquid is set in motion, not by the whole speed v *of* this, but only by the fraction

$$\left(1 - \frac{1}{n^2}\right)v,$$

so that the light propagates within the liquid with the overall speed

$$\frac{c}{n} + \left(1 - \frac{1}{n^2}\right)v,$$

The experiments made by Fizeau and repeated in more recent times by Michelson and Morley confirmed the accuracy of Fresnel's prediction. By Fresnel himself and by the greatest scholars of theoretical optics an attempt was made to justify, on the basis of the various hypotheses on the nature of light, the previous formula. But the simplest explanation comes from Einstein's theory. The speed of light $\frac{c}{n}$ is actually composed with the velocity of the medium v, but the composition occurs according to the formula of the new kinematics, which, given the relevant value of $\frac{c}{n}$, differs considerably from the classical formula, and leads to the above result. The drag of the light waves in the liquid in motion, which seems *partial* when we adhere to Galileo's kinematics, is, in fact, *total* for Einstein.

This very simple deduction (of which I have here reproduced only the general outline) of the experimental law of propagation of light in a body in motion, starting from the hypothesis of the constancy of the velocity of light in vacuum, certainly constitutes a brilliant confirmation of Einstein's theory. It is to be hoped that some other immediate consequence of the two principles which form the basis of that theory may be submitted to investigation by experiment. The attempts made, with admirable experimental skill, by Kaufmann and others to verify more remote consequences of the theory, leave some doubt as to the interpretation that was made of them; since the confirmation or denial of complex theoretical results involves not only the two principles of which we have spoken several times, but all the other more or less explicit hypotheses that must be added to those principles in order to obtain the said results.

Summary

In order that the reader may find his way through this paper, where the desire for clarity has sometimes forced us to linger over details, it is appropriate that I summarize the main points.

1. The principle of relativity in its vague form is not disproved by any fact and corresponds to a need of our spirit. It is, on the other hand, one of those *elastic* truths, susceptible of assuming different aspects according to the way in which one tries to specify them, and therefore destined to survive the cataclysms that from time to time strike the most concrete scientific theories.
2. In the study of optical phenomena, the said principle was specified by Einstein with the further hypothesis that the phenomena themselves satisfy it when only the material bodies participating in the experiment are taken into account, leaving out the aether regarded as a representation devoid of reality.
3. For the speed of propagation of light in vacuum there are several hypotheses that agree with the principle of relativity. The most plausible are: the ballistic hypothesis, according to which the speed of light is composite (in Galileo's meaning) with the speed of the source and the observer; Einstein's hypothesis, according to which the speed of light is a universal constant, whatever the motion of the source or the observer. The first hypothesis conforms to classical kinematics, but does not seem to be confirmed by the facts; the second hypothesis is incompatible with said kinematics, but the consequences to which it gives rise seem to agree with experiment.
4. Einstein's hypothesis forces us to modify profoundly our ideas about lengths, times, contemporaneity, and to construct a kinematics more complex than the ordinary one. The latter would be valid only in first approximation for speeds small compared to that of light.

The complications brought by Einstein's views to the most elementary concepts themselves, seem so serious to several distinguished physicists as to make them distrust the new theory. And this distrust may be providential, since the progress of science requires that no complication should be accepted until it has been shown to be necessary, or at least capable of providing a synthetic vision of wider horizons.

On the other hand, it is legitimate to ask whether the agnostic position in front of the aether, with which Einstein specifies the principle of relativity, satisfies our physical intuition.

Certainly the hypothesis that the configuration properties, the characteristics and the physical constants of the aether are rigorously the same throughout the universe, leads us to return again to absolute space, from which we try to free ourselves. But in the face of that hypothesis of absolute homogeneity, which has a purely mathematical character, physics suggests to us a concept of statistical regularity, analogous to that which the kinetic theory admits in gases. Could not the characteristics of the aether vary, even if within narrow limits, as time and position in space change? The laws of a vibratory propagation in such a medium would naturally have an average value, and could give rise to deviations, whose absolute values would increase with the extension of the field of observation.

A view based on such hypotheses would still satisfy a principle of relativity, not in Einstein's sense, but analogous to that which governs the acoustic phenomena in our atmosphere. Relativity would indeed be more complete, since the same universal constant (speed of light) that in Einstein's theory survived the shipwreck of the absolute, would be replaced here by an average value. Classical optics would be obtained through a passage to the limit, extending to the entire universe the supposed homogeneity of the medium in an infinitesimal space. But precisely this passage to the limit would resurrect the absolute, which more or less explicitly accompanies every hypothesis of homogeneity.

Will it be possible to construct a satisfactory statistical theory of optics and electromagnetism? This is a question which will wait for a long time to be answered, but which is suggested by the evolution of scientific thought. The domain of the great and simple natural laws attributing multiple consequences to a single cause is less extensive than it seemed a century ago. Our interest is now attracted towards the evaluation of the effects resulting from the concurrence of innumerable and minute causes. The methods of research in the physical sciences and in the social sciences are tending more and more to come closer to each other.

The New Mechanics

Max Abraham

What is it and why are we dealing with a "new mechanics"? Does not the old classical Galilean-Newtonian mechanics correctly describe the motions of bodies, both terrestrial and celestial? Certainly the principles of classical mechanics make it possible to describe the motions of masses under the influence of their mutual gravitation. But are they sufficient if the forces of electricity and magnetism, of light and heat, also come into play?

Mechanics has always claimed the right to be able to give an explanation to all these forces. Not only should its geometrical-kinematic concepts be the basis of the whole of physics, but all natural phenomena should be led back, in the last resort, to processes of motion. This tendency lies at the basis of the kinetic theory of gases and Maxwell was well aware of it since his sketch of the dynamic theory of electromagnetic fields. Maxwell was able to show that the electromotive and ponderomotive forces exerted between two electric circuits obey the equations of Lagrange's mechanics. Following Maxwell, J. J. Thomson and H. Helmholtz respectively elevated Lagrange's equations and the principle of least action to fundamental principles of all physics.

Hertz, finally, in his *Principles of mechanics,* proposed to lead all the forces present in nature back to the inertia of moving masses. Even where we do not perceive matter, to fill the space there must be masses that are not seen which, coupled with each other, transfer energy from one body to another. Apparent forces at a distance should therefore always be understood as actions in contact between hidden masses.

With the posthumous works of H. Hertz, the phase of research whose objective was to subordinate the whole of theoretical physics to the principles of mechanics without essentially modifying its content, came to an end. Hertz's mechanics remained in a programmatic state and the direction indicated by it was not pursued further.

M. Abraham, *Die neue Mechanik*, "Scientia", 15, 1914, pp. 8–27.

With the need to banish actions at a distance from mechanics, the influence of Faraday and Maxwell's concepts on the electromagnetic field is evident. They triumphed thanks to Hertz himself, and would have governed the following phases of development. According to Maxwell's theory the electric and magnetic forces propagate from point to point with a finite velocity that, in vacuum, is $c = 3.10^{10}$ cm/sec. The luminous and thermal radiations are simply a particular type of electromagnetic waves. It is not necessary to imagine the "aether" as the propagation medium of the electromagnetic fields, or to think of space solely as provided with physical characteristics even where there is no ponderable matter. It is instead sufficient to know the equations that describe the propagation of the electromagnetic forces in space or in the "aether". However, it is fundamental that a definite positive amount of "energy" is attributed to every electric or magnetic field. Modifications of the field are accompanied by an "energy current" that, according to a Poynting theorem, takes place orthogonally with respect to the magnetic and electric force.

In the case of plane waves, the Poynting vector of the energy current becomes identical with the ray-vector of optics. According to Faraday and Maxwell, the transmission of forces from one body to another takes place by means of certain fictitious "tensions", that is, a "pull" along the lines of electric and magnetic forces and a "pressure" perpendicular to them.

What effects occur when light is propagated from a body A and subsequently absorbed by another body B? The energy emitted by A flows along the light rays with a velocity c, until it meets B. At the moment of emission A undergoes a reaction; at the moment of absorption B feels a blow of equal magnitude but which is exerted in the opposite direction. The existence of a pressure exerted by light has been demonstrated both experimentally and theoretically. Is it possible to reconcile the forces of light pressure with the old mechanics?

Newton's third axiom postulates that an action corresponds to an equal and opposite reaction. If they are spatially separated bodies, a force can correspond to a simultaneous reaction only when the propagation of forces occurs instantaneously. With a finite speed of propagation of the forces, the principle of action and reaction in its classical form is incompatible since not only the equality, but also the contemporaneity of *actio* and *reactio* is necessary. Indeed, one must take into account the fact that force, as well as energy, remains latent for a certain period of time.

It is possible to describe the phenomena in the most immediately perceptible way by assigning to light, as to all electromagnetic fields, an "electromagnetic momentum". During the emission of light, this momentum is subtracted from body A and sent into space; in the absorption process, it is reacquired by body B. It is then possible to consider the theorem of the conservation of the momentum valid in the sense that the sum of the momenta of the matter and of the field remains constant. During the emission of the light, the material momentum of body A is transformed into an electromagnetic momentum; the force that the light exerts during the emission on body A corresponds to this. During the absorption process, the electromagnetic momentum of light is transformed into the material momentum of body B. The opposing force that the light exerts on body B which is absorbing energy corresponds to this. A general expression for the electromagnetic momentum contained in the unitary volume has also been found, which is equal to the Poynting energy flux divided by the square of the speed of light. Each energy flux in the field thus produces an electromagnetic pulse. This interpretation corresponds specifically to H. A. Lorentz's representation of Maxwell's theory.

This generalization of the impulse theorem should become decisive for an important problem of the new mechanics, the dynamics of "electrons". These are those particles of negative electric charge which, coming out of cathodic tubes, form the so-called cathode rays. As A. Schuster hypothesized and W. Kaufmann demonstrated, the deflection of these rays by a magnetic or electric field can be explained by attributing to them, besides a negative charge, an inertial mass. However, the mass of the electrons proved to be much smaller (approximately in a ratio of 1:2000) than the mass of the hydrogen atom; their velocity is equal to $c/10$ in cathode rays. Even faster, almost equal to the speed of light, electrons in motion are identified in the so-called β-rays of radioactive bodies. The question now arises: is the second axiom of Newtonian mechanics still valid for these velocities? Is the acceleration of the particle still equal to the initial force divided by a mass of the particle independent of its velocity? Experiment (Kaufmann, 1901) showed that this is not so. During the experiment, the inertial mass in question, i.e. the ratio between the deviating force of the electromagnetic field and the acceleration resulting from the deviation, increases with increasing velocity. In electron dynamics therefore, Newton's second axiom is no longer valid.

Within the new mechanics the necessity arose to adequately extend the concept of inertial mass. The solution given is based on the extension of the impulse theorem mentioned above. The electron generates an electric field with its charge and a magnetic field with the motion of the charge. Around it flows a current of electromagnetic energy possessing an electromagnetic momentum. The total electromagnetic impulse of the electron (G) is a function of the velocity (q). Only two cases must be distinguished: acceleration in the direction of motion and acceleration orthogonal to it.

In the first case, the derivative of the impulse with respect to the velocity ($\frac{dG}{dq}$) gives the ratio between the variation of the impulse, that is the force, and the variation of the velocity, that is the acceleration; this is the so-called "longitudinal mass". If the acceleration is orthogonal with respect to the trajectory, impulse and velocity instead remain constant in modulus. Only their direction changes, in this case, from the ratio of the variations of the impulse-vector and of the velocity-vector the expression $\frac{G}{q}$ finds validity.

This "transverse mass" is equal to the longitudinal mass only in the case of small velocities, where G is proportional to q. As regards velocities of the order of that of light, as follows from Kaufmann's deflection experiments, G is no longer proportional to q; consequently, the longitudinal mass is, in this case, different from the transverse mass. If the force is positioned in a direction oblique to the velocity, then the above acceleration will not be parallel to the force. For as a consequence of the difference between the longitudinal and transverse masses, the two components of the acceleration, longitudinal and transverse, are in a different ratio than those of the respective force components.

Physicists of the old school, faced with such a subversion of the concept of mass, shook their heads thoughtfully. They had in fact to admit that classical mechanics was valid only for small speeds in comparison to the speed of light. They could think to differentiate the electromagnetic mass of the electron from that of the material atom and to consider this latter mass, according to classical mechanics, as constant. The development of physics goes in the opposite direction. It was tried to consider the atom as an agglomerate of electrons; the electrons contained in the atom give a contribution to the atomic mass and condition its dependence on the velocity. In this way the new mechanics also drags matter within its own sphere.

On the other hand, it is noteworthy that very definite principles of analytic mechanics—Lagrange's equations and the principle of least action—retain their validity even in the new mechanics, if one generalizes the expressions for the Lagrange function and for the action in an appropriate way. However, this is not the place to tackle this subject.

The problems faced up to now can be treated without any doubt on the basis of electromagnetic field theory. The greatest difficulties in field theory arise from those questions that have to do with the relativity theorem. In classical mechanics there is this theorem: in a material system that is in a state of uniform translation, the relative motions take place exactly as in the same system in a condition of quiet. An observer belonging to the system, observing the processes from within it, cannot therefore observe the presence of that translation motion. Now, does such a theorem hold also in a system of electrons? If one ascribes a real existence to the electromagnetic field in space, or if one ascribes to it the more or less substantial aether as a vector, one might suppose from principle that a theorem of relativity is valid only if the aether moves together with the electrons.

On the contrary, Lorentz's electrodynamics, on which the electron mechanics briefly described above depends, assumes that the aether does not take part in the motion of the electrons. How can it be then that the motion of a system is not detectable by observers moving along with the events taking place in the system? How can it be that, for example, optical experiments on light from terrestrial light sources leave no trace of an effect of terrestrial motion? This problem was dealt with by H. A. Lorentz in a series of papers (1892–1904).

The concept of "local time" plays an essential role in the above works. Let us consider a system in uniform and rectilinear motion in the aether in a state of rest. At the single points of the system there are stations which have clocks all equal to each other. These clocks are regulated by a light signal, where only the constant speed of light is taken into account but not the unknown motion of the system. The time provided by one of the clocks is called the "local time" of the station in question. Lorentz now showed that, with reference to "local time", electromagnetic and optical processes take place in the system in motion exactly as if the system were at rest. He initially considered only first-order quantities with respect to the ratio $\beta = \frac{q}{c}$ between the velocity of the system and the speed of light (for the Earth it is $\beta = 10^{-4}$). Later he generalized the result, considering also second order quantities in β.

In this regard, it concerned in particular the explanation of the experiment conducted by Michelson in order to observe the effect of the earth's motion $\sqrt{1 - \beta^2}$ through the observation of interference.

Starting from a single source of monochromatic light, Michelson made two rays coming from it interfere, one of which with a propagation parallel to the motion of the Earth, the other with a propagation orthogonal to it: all this with the expectation of detecting the influence of the Earth's motion on the interference lines. The result was, however, negative. We obtain a schematic theory of such interference experiments if we observe a ray of light transmitted from a point O of a system to another point P from where, by means of a mirror, it is reflected back to the point O. If the system is at rest, the geometric locus of the points P, which correspond to equal optical paths $OP + PO$, is evidently a sphere. If, instead, the system is in uniform motion, it is necessary to evaluate the absolute optical paths to be associated with the relative optical paths OP and PO passing through the moving system. As geometric locus of the points P that correspond to equal absolute optical paths, a rotation ellipsoid is identified, the so-called "Heaviside-Ellipsoid", which has a ratio between the axes equal to, and is flattened, in the direction of the motion of the system. Since the difference between the absolute optical paths of the two light rays is decisive for their interference, it must be deduced from the negative result of Michelson's experiment that the absolute optical paths of the two rays remain equal in a moving system, if they were equal in the case of rest: which means that the points P, which were on the sphere in a condition of rest, form the "Heaviside-Ellipsoid".

From this reasoning the "contraction hypothesis" of Fitzgerald and H. A. Lorentz flows: all bodies placed in motion, should contract parallel to the direction of motion, since all segments parallel to the direction of motion shorten according to the ratio $\sqrt{1 - \beta^2}$:

1. Lorentz tried to make this apparently singular hypothesis plausible by assuming that the molecular forces, which determine the shape of bodies, are electrical in nature. Then, as he showed, the change in the electric and therefore also molecular forces determined by the motion of the system would produce precisely the observed change in shape. If now in the system thus contracted an optical or electromagnetic time regulation still applies, then the relativity theorem is also valid in relation to quantities of second order or any order in β. To the observer moving with the system and following the processes of the system itself the uniform motion of the system remains hidden. In this way, Lorentz succeeded in reconciling the negative outcome of the experiments conducted up to then on the influence of the Earth's motion with the theory of the electromagnetic field.

Although Lorentz still expounded his results in the typical language of classical kinematics, it was precisely from these that he felt the need for a break with the concepts of time and space of classical mechanics if the relativity theorem was to be applied to electromagnetic fields. If the motion produces a contraction of all bodies, then the rulers used in the measurement of the lengths are also shortened. The measurements effected in a moving system cannot then lead to the knowledge of that "absolute" form of the bodies with which classical geometry deals. On the other hand, an electromagnetic clock in a moving system—and after Lorentz's observations on the constitution of matter, every clock behaves like an electromagnetic clock—will always show the corresponding "local time" and not the "absolute" time of classical kinematics. The question had to be asked whether, with these premises, the geometrical and kinematic knowledge transmitted up to then still had validity.

To this question Einstein gave a negative answer. His theory formulated in 1905, today known as the "theory of relativity" is based on two postulates: the first requires the equivalence of systems moving in uniform motion with respect to each other (*Postulate of relativity*). The second asserts that in each of these systems the transmission of light in space occurs in every direction and with the same speed (*Principle of the constancy of the speed of light*). Relying on these postulates, Einstein gives relative definitions of length and time. Absolute definitions of these quantities are no longer possible with the theory of relativity since, according to the first postulate, there is no means to determine whether the system, within which the measurements are made, is at rest or in uniform motion.

The theory of relativity, to which H. Minkowski (1908) later gave an appropriate mathematical expression, awakened the attention of other circles to the new mechanics. The present revolution of the traditional concepts of kinematics and dynamics surprised those who had not followed the historical development of these problems recalled so far. The apparent universality of the solution given to the problem of space–time met the philosophical need of that historical moment which was in search of a uniform and all-encompassing knowledge. So the young who studied mathematics and physics were thrilled by the theory of relativity, which in that era influenced by it crowded the lecture rooms of the universities. The physicists of the older generation, on the other hand, whose training had been influenced by Mach and Kirchhoff, looked with skepticism at the new young daredevils who dared to overturn the foundations of traditional physical measurements, on the basis of a few experiments debated by experts in the field. Some might paraphrase Wallenstein:

> Youth is very ready with words, [...].
>
> in its hot head it boldly takes the measure of things.

Although many interesting articles in this journal have already dealt with the theory of relativity, its discussion in connection with this article may not seem superfluous.

How do the interpretations of the theory of relativity differ from those of Lorentz's theory? The claims of the two theories are essentially identical. From the point of view of an observer taking part in the motion of the system, Einstein's rulers present the Lorentz contraction, and Einstein's clocks indicate Lorentz's "local time". Relativistic dynamics also agrees in all respects with Lorentz dynamics. What differs is only the starting point of the treatments. Lorentz bases his theory of the electromagnetic field on the hypothesis that matter is electrically constituted; for him, the theorem of relativity arises as a consequence of these premises. Einstein, instead, sets as fundamental postulate the property of relativity: the hypothesis of a universal electromagnetism becomes superfluous. Naturally, the second postulate brings with it the fact that the speed of light enters into the laws of dynamics and into all the laws of physics. However, it is conceivable that such a relationship may exist without the fact that all processes must be electromagnetic in nature. In this way the interpretation given by the theory of relativity turns out to be the most general. However, according to this theory, no change of state can propagate in vacuum with any speed other than that of light. Otherwise they could be used as transmission signals in a system in motion, thus arriving at a new definition of time and length, different from the one given by Einstein. Starting from the difference in the operation of two clocks, the first optical and the other regulated by means of the new change of state, the motion of the system could be calculated. Since the first postulate does not allow it, according to the theory of relativity, all forces in space, including the force of gravity, propagate with the speed of light. Propagation and motion at a speed faster than the speed of light do not exist.

As a consequence of the fact that instantaneous transmission of physical states is excluded, it follows that it is impossible to transfer the concept of a "rigid body" from classical mechanics to the theory of relativity. The rigid body of classical mechanics has six degrees of freedom. If one prescribes the motion of three of its points (whose distance obviously remains constant), the motion of all the remaining points of the body is determined. In relativistic mechanics it is not possible to establish such a relationship, since, as M. Laue has observed, this would mean that the condition of motion would spread instantaneously through the body. Since, on the other hand, the process of motion needs time to propagate from one point to another, it must be possible to establish the motion of any large number of points. This means that the number of degrees of freedom is infinite. The often so convenient idealization of classical mechanics, which replaced solid bodies with rigid ones, is not accepted within the theory of relativity. One must base the mechanics of extended bodies on the relativistic theory of elasticity, as developed by G. Herglotz.

Some supporters of the theory of relativity derive from the first postulate the uselessness of the idea of the "aether" as a medium that fills space. Undoubtedly, as a consequence of this postulate, the "aether" is imperceptible in the condition of uniform rectilinear motion. On the other hand, as P. Ehrenfest pointed out, the second postulate on the constancy of the speed of light is not fully comprehensible without the use of wave theory. (In the emission theory of light, the speed of light itself depends on that of the light source; in it the second postulate is not satisfied.) The second postulate testifies in favor of deriving relativity theory from field theories. Radical relativists, the enemies of the "aether" would like to disprove such an origin.

With respect to this, E. Wiechert makes the following observation: if the correctness of all the assertions of the theory of relativity were proved, above all the existence of an upper limit for all velocities, all physicists would certainly be pushed to formulate a field theory for which every limit velocity would find its own explanation as the velocity of propagation of energies in a medium that fills space. Wiechert tries in this way to reconstruct the "aether" starting from relativistic ideas.

Moreover, in accordance with the historical origins of the theory of relativity, many traits of the Maxwell-Lorentz theory can be found there. The aforesaid concepts of field energy, tensions, energy current and field impulse are present here. In it there is also the relationship between the last two vectors mentioned, formulated through the following theorem: the momentum contained in the unit volume is equal to the energy current, divided by the square of the speed of light. This theorem, which M. Laue calls the "energy inertia theorem", is considered valid even by M. Planck for energy currents of various kinds. It follows an important generalization of a relationship already present in field theory between the inertial mass of a body and its energy. In a dynamics thus summarily sketched, the electromagnetic mass of a slowly moving electron is proportional to its electrostatic energy. Similarly, a cavity filled with thermal radiation possesses, according to F. Hasenöhrl and K. Mosengeil, an inertial mass that is proportional to the total energy of the radiation. Taking into account the theorem of the inertia of energy, if we also attribute a momentum to the energy currents, which are produced by the voltages inside the electron, or in the wall of the cavity, the relationship is simplified and proves to be

$$m = \frac{E}{c^2}$$

This applies to all forms of energy.

The inertia of matter is, from now on, determined by its energy content. To each loss of energy, e.g. through the emission of heat, corresponds a decrease of the inertial mass. Of course, considering the smallness of the factor c^2, the decrease of the mass,[1] corresponding to the thermal content of normal chemical reactions, is extremely small. The overall energy of the molecules, on the other hand, calculated from that formula using the value of its mass, is much larger than that which is at issue in these chemical reactions. This energy is assumed to be located within the atom. However it is known that in radioactive transformations, the atom gives up quantities of energy an order of magnitude much greater than that which develops in chemical reactions. In these transformations a decrease in mass also takes place. However, almost the entire mass of the transformed atom is found in the products of the transformation, to which the helium atoms emitted in the form of α-rays also belong. That minimal part of mass that is lost with the kinetic energy of the propagated rays and of the yielded heat is very small (about 10^{-4}). This does not mean that it is negligible. If it were possible to formulate exactly the balance of the mass in radioactive transformations up to an order of magnitude equal to 10^{-4}, it would be possible to verify every theoretical relationship between mass and energy.

As we have seen, the theorem of the inertia of mass does not belong to the theory of relativity. Instead, it has its roots in field theory regarding the transmission of energy and forces, which found their fullest development only through the influence of relativistic ideas.

Let us return to the contrast between field theory and the theory of relativity. Field theory provides a self-consistent picture of the world that encompasses a wide range of experience. In contrast, the theory of relativity is incomplete. It limits its assertions only to the case of uniform rectilinear motion, whereas electromagnetic field theory also refers to accelerated and curvilinear motion. Such motions of the

---

[1] This is an oversight of the text. What is small is not c^2, but $1/c^2$.

(Silvio Bergia is gratefully acknowledged.).

electrons provoke electromagnetic waves that carry with them energy: in this way the emission of light is explained. In consequence of this, there is no equivalence between the uniform and rectilinear motion of the electrons on the one hand and the accelerated and curvilinear motion on the other, because this latter motion does not have any loss of energy as a consequence.

The field, by subtracting energy from the electron, acts on every motion of the electron that deviates from the uniform rectilinear one. The radical relativists are therefore not right in presenting the idea of the field and its vector as superfluous. As yet it is not possible to emancipate the theory of relativity from that of fields, being bound to the latter by the second postulate.

The origin for every radical relativist tendency is evidently the first postulate, which invites us to extend our study also to motions of a different nature. If it were possible to extend this postulate also to non-uniform and rotating motions of a system, i.e., to demonstrate the equivalence of a system in such a state of motion with one at rest, the "aether" would become in fact useless. However, from the fact that the rotation of a mechanical system is shown by centrifugal forces, it follows that two rotating systems are not equivalent with respect to each other.

We recognize in Einstein's 1905 theory of relativity a compromise between two heterogeneous concepts. The more rigid relativist idea strives to describe the relative motion of matter as the only decisive element while the field and its vector are superfluous. From this the first postulate arises, while the second is borrowed from field theory. This internal discord contains in itself the germ of ruin. That compromise could last only so long as it dealt solely with uniform rectilinear motion and chose very carefully the setting of the phenomena to be treated. Any attempt to extend the theory had to overcome this unnatural pact.

The crisis of the theory of relativity began when it also aimed to include the force of gravity in its observations. We spoke before about the inertia of energy and the possibility of verifying it through an exact energy and mass balance of radioactive transformations. Now, however, the chemical determination of mass with a balance does not refer to the inertial mass but to the weight. However, since Galilei showed that the force of gravity imparts the same acceleration to all bodies, the proportionality of weight and inertia is a recognized law. If the inertial mass were not always equal to the gravitational mass, the force acting on the masses at the surface of the Earth would have different directions for chemically different bodies, since it is the result of the Earth's attraction, which is proportional to the gravitational mass, and the centrifugal force, which is proportional to the inertial mass. That this is not so, but rather that the direction of the resultant force acting on the masses, i.e. the vertical, is the same for all bodies, has been shown by B. Eötvös by quite precise investigations. If we also ascribe inertia to energy, we must also give it gravity, according to the relation

$$m\,c^2 = E$$

If we consider this *relationship between gravitational mass and energy*, the laws of conservation of energy and *conservation of weight* converge into a single law.

Now, however, the energy of a body changes with displacement in a gravitational field. It increases, for example, when a body is lifted since a work opposing the force of gravity is produced. To this change of energy E must correspond, as a consequence of the above expression, a change of the mass m or of the speed of light c in a gravitational field.

The latter hypothesis—*dependence of the speed of light on the gravitational potential*—was formulated in 1911 by Einstein who with this gave a clean break to one of the roots of his first theory of relativity.

Subsequently, in 1912, I developed a theory of the gravitational field which is based on this line of thought. As has been mentioned above, there is reason to consider that gravitation propagates with the speed of light. It is B.W. Ritz who explains that astronomical facts do not stand in the way, ("Scientia," No. X-2, vol. V, 1909). However, according to a general theorem of wave theory, in a medium in which waves propagate with definite velocity, only two types of waves are possible: transverse or longitudinal. For the former field equations of the type of Maxwell's equations are valid; admitting transversal waves would lead therefore to an electromagnetic gravitational theory. However, some difficulties arise along the path to reach it, as already observed by Maxwell, in consequence of the different direction that the forces can have, that is to say the attraction between masses with respect to the repulsion between equal charges. The energy of the gravitational field becomes negative: this involves, as for example B. W. Ritz stresses in the article just cited, an instability of the equilibrium in such a field theory. Nevertheless, after Maxwell many well-known researchers dealt with the electromagnetic theory of gravitation, but without success. The electromagnetic model seemed to be too limited to encompass even gravitation. If a satisfactory field theory of gravitation is to be arrived at, the hypothesis of a universal electromagnetism must be abandoned. The above reasoning, however, leaves no other way out other than hypothesizing a propagation of gravitation by *longitudinal waves*. Thus I applied the famous wave equation to the gravitational potential; in static fields it reduces to Poisson's equation of the traditional potential theory.

In the new field theory of gravitation, the part of energy introduced into the field becomes positive; the equilibrium of masses at rest will become stable. The transmission of forces occurs by means of voltages, those of energy by means of a current of energy, which possesses a momentum, in accordance with the theorem of the inertia of energy. Therefore it is possible to realize all the essential teachings of the electromagnetic field theory, even if the electromagnetic model is completed in the way sketched above. The electromagnetic field depends on six quantities, that is to say the components of the electric and magnetic vector, while the gravitational field, instead, depends on four quantities, that is to say the partial derivatives of the gravitational potential with respect to the coordinates and to time. In the first case there are six, in the second one instead there are four quantities that determine the energy, the energy current and the voltages of the field.

If one were to combine Einstein's idea of the link between the speed of light and gravitational potential in this way, the variations in the value of the speed of light, which propagates in those longitudinal waves, would be similar to the way pressure variations propagate in sound waves, and therefore also to those of the speed of sound.

The energy gravity theorem can therefore be safely incorporated into the theory.

G. Mie gave a further modification of the theory in accordance with Einstein's 1905 theory of relativity. In order to keep the postulate of the constancy of the speed of light valid, this physicist made mass depend on gravitational potential instead of on this constancy. Since not the sum, but the difference of potential (or electric) energy and kinetic (or magnetic) energy remains unchanged with respect to the space–time transformations of the theory of relativity, he found himself forced to pose the gravity of a body as proportional to this difference. As a consequence of this a positive weight would correspond to the potential (or electric) energy, while to the kinetic (or magnetic) energy a negative weight would correspond. During a mechanical or electromagnetic oscillation in a closed system, in the course of which transformations of energy periodically take place from the potential form to the kinetic one, or from the electric form to the magnetic one or vice versa, the weight of the system would undergo minimal, almost imperceptible periodic modifications. Moreover, in this relativistic theory of gravitation, the balance of the weight in the radioactive processes would not be in conformity with the balance of the energy in the radioactive processes. If from an element A an element B derives which contains a minimal quantity of kinetic or magnetic energy, an increase in weight would correspond to the reduction of the energy, since this quantity of energy must be related to a negative weight. In the context of this theory, therefore, the law of conservation of weight in a closed system would not be strictly valid. Since, on the other hand, the inertial mass remains proportional to energy, according to Mie the identity between gravitational mass and inertial mass is also no longer exact. In my opinion, instead, the theorems on the conservation of weight and on the identity of gravitational and inertial mass seem proved at least as exactly as the constancy of the speed of light. In agreement with Einstein, I too therefore prefer to sacrifice the latter.

In a recently published essay G. Nordström tries to refine the gravitational field theory in such a way that the proportionality between gravitational and inertial masses is preserved in as large an extension as possible, without, however, abandoning the postulate of the constancy of the speed of light. This also applies to systems that are at rest or in stationary motion. However, Nordström fails to show that in radiative transformations the weight changes in a way proportional to inertia.

Moreover, Nordström's theory is forced to consider the length of each ruler and the motion of each clock as dependent on the gravitational potential. In the gravitational field, within which the potential changes its value from place to place, the indications of such rulers and clocks would not fit within the relativistic and universal space–time scheme, for which the postulate of the constancy of the speed of light has validity. It remains unresolved whether such a departure of the local measurement of space–time from the universal and optical definition of space–time in the meaning given by the theory of relativity of 1905 can still be accepted.

Einstein himself dealt with the theory of gravitation, starting from premises that belonged partly to classical mechanics and partly to the theory of relativity of 1905, although he abandoned its second postulate. He even raised hopes for a new and more general theory of relativity that would also include gravitation, the outline of which is contained in *Draft of a theory of generalized relativity and of a theory of gravitation* (*Entwurf einer verallgemeinerten Relativitätstheorie und einer Theorie der Gravitation*, A. Einstein and M. Grossmann, B. G. Teubner, Leipzig, 1913).

The authors introduce an arbitrary space–time transformation. The gravitational field should depend on the ten coefficients which define space and time in the infinitesimal. The static field should influence only the value of the speed of light and consequently also the movement of the clocks; the dynamic field should, moreover, cause dilations and deformations in the grid of the coordinates. This principle of a generalization of the theory of relativity is not without greatness. The authors of the "Entwurf", with the help of the method of the "absolute differential calculus" developed by Ricci and Levi–Civita, succeed also in giving a form to the fundamental dynamical and electromagnetic expressions that, at least in the infinitesimal, is enough for what relativity demands with regard to that general transformation. However, the same gravitational field, whose unification with the theory of relativity seems to be the main purpose of the "Entwurf", does not fit the point of departure. The differential equations of the gravitational field obtained by the authors are not invariant with respect to that general space–time transformation, which means that a system of mutually attracting masses in non-uniform or rotational motion is not in general equivalent to a system at rest.

The fact that there is no relativity with regard to rotational motions is well known; if it were to the contrary we would not be able to deduce the rotation of the Earth from observations about its shape and its gravitational field. The problem of "absolute rotation" from the point of view of classical mechanics has already been debated at length. Most authors consider the fundamental equations of dynamics as non invariant with respect to rotational motions. This interpretation is justified only if one considers the force field as already given. In contrast, Giorgi has shown how one can give a relativistic form even to Newton's equations of motion. In fact, an observer bound to a precise place on the Earth's surface should be able to calculate all the forces acting on the masses, including also the centrifugal and Coriolis forces, as "gravitational forces". In this way he can leave unchanged the second axiom of Newton's mechanics, in spite of the rotation of his own reference system. The difficulty, however, moves only from the field of dynamics to the theory of gravitation and in a corresponding way the "Entwurf" of Einstein and Grossmann also proceeds. In fact, once the observations from the different stations have been collected, that is to say once the gravitational field of the Earth has been determined, the fact is shown that this field does not correspond to the one that the Earth should give rise to in a condition of quiet according to the Newtonian law of attraction. According to this theorem, in fact, the sources of the gravitational field are found only inside the masses that exert the attraction. The apparent gravitational field of the rotating Earth, on the contrary, possesses a component (the centrifugal force field) whose origin depends on the speed of revolution of the Earth. If therefore the further forces of the rotation are added to the "gravitational forces", the dynamics becomes relativistic, but the gravitational field of a rotating system does not remain equivalent to that of a system that does not rotate. It is therefore wrong to look for a theory of gravitation corresponding to the general relativistic scheme. Assuming such a theory were accepted from a mathematical point of view, it would not be correct from the physical point of view.

Faced with the stumbling block posed by gravitation, therefore, every theory of relativity, whether the restricted theory of 1905 or the general theory of 1913, fails. Relativistic ideas are clearly not broad enough to provide a complete picture of the world.

However, the theory of relativity has historical merit in its critique of the concepts of space and time. For it has taught us that these concepts depend on the representations we construct for ourselves of the behavior of the rulers and clocks that serve for the measurement of lengths and intervals of time, and that these, like the former, are subject to change. Hence the theory of relativity is assured a funeral with all honors.

Whatever the fate of the theory of relativity, the new mechanics will continue to prosper, since it does not formulate a rigid system of axioms, but sees as its task that of keeping mechanics in contact with the other disciplines of physics. In this way, the alternating fortunes of the other theories of physics will also influence mechanics, which in turn will regulate and act decisively in the construction of a *Weltanschauung* proper to physics.

On the Problem of Relativity

Albert Einstein

Since two high ranking experts have expressed misgivings about the theory of relativity in this journal, readers might appreciate it if a proponent of the new theoretical course would also state his views. This will be done briefly.

Today we must distinguish well two theoretical systems which are both called "theory of relativity". The first of these, which we shall call "theory of relativity in the narrower sense", is based on a considerable complex of experiments and is today accepted by the majority of theoretical physicists as the simplest theoretical expression of the experiments. The second (called by us "theory of relativity in the broader sense") finds almost no basis in the experiments of physics so far. Most fellow specialists regard this second system skeptically or unfavorably. It must be made clear from the outset that one may well be an advocate of the theory of relativity in the narrow sense without accepting the validity of the theory of relativity in the broader sense. We shall therefore treat the two theories separately.

The Theory of Relativity in the Narrow Sense

It is well known that the equations of mechanics, founded by Galileo and Newton, are not valid for any system of coordinates in motion, if for the description of motions we admit only central forces that answer to the laws of the equality between *actio* and *reactio*. But if we refer motion to a system K, making Newton's equations valid in the way mentioned, that coordinate system is not the *only one* in relation to which those laws of mechanics are valid. Any coordinate system K', oriented at will in space that is in uniform translation motion with respect to K, has instead the property that

A. Einstein, *Zum Relativitäts-problem*, "Scientia", 15, 1914, pp. 337–348.

in relation to it the same laws of motion hold. The assumption of the equality of all coordinate systems K, K', etc. for the formulation of the laws of motion, indeed of all general laws of physics, we call the "principle of relativity" (in the narrow sense).

As long as it was believed that classical mechanics had to constitute the base of the theoretical description of all processes, it was not possible to doubt the validity of that principle of relativity. But even leaving this aside, even from the point of view of experiment it is difficult to doubt the validity of that principle. If it were not valid, the natural processes of a reference system that is motionless with respect to the Earth should appear to us to be influenced by the motion (velocity) of annual revolution of the Earth around the Sun; the Earth's observation spaces should behave in a physically anisotropic way because of the existence of such motion. Despite the most laborious research physicists have never been able to observe such an apparent anisotropy.

The principle of relativity is therefore as old as mechanics and, from the point of view of experiment, no one could ever have doubted its validity. That, however, it has been questioned and still is questioned today is due to the circumstance that the principle of relativity is in apparent incompatibility with the electrodynamics of Maxwell and Lorentz. Anyone capable of judging the compactness of that theory, the small number of assumptions underlying it, and its merits for the theoretical description of experiments in the fields of electrodynamics and optics, can hardly reject the feeling that the foundations of that theory are to be considered equally definitive as the equations of classical mechanics. Nor has it been possible to place side by side with that theory another that is even partially able to compete with it.

It is easy to point out the reason for the apparent incompatibility between the electrodynamics of Maxwell and Lorentz and the principle of relativity. The equations of that theory can be valid relative to the coordinate system K. By this it is said that every ray of light propagates relatively to K in vacuum with a certain velocity c independent of the direction of propagation and of the state of motion of the source of light. This statement will be indicated henceforth as the "principle of constancy of the velocity of light". When such a ray is considered by an observer in motion with respect to K, the velocity of propagation of the ray appears, from the point of view of this observer, generally different from c. If the ray propagates e.g., with the velocity c in the direction of the positive axis x of K, and if our observer moves with the constant velocity v in the same direction, we believe we can immediately state that the velocity of propagation of the ray, judged from the observer's point of view, is $c - v$. With respect to the observer, that is to say, with respect to a reference system K', the principle of the constancy of the velocity of light seems not to be valid. We are thus faced with an apparent contradiction to the principle of relativity.

As everyone knows, a precise analysis of the physical content of our spatial and temporal assumptions has had the result that the above contradiction is only apparent, since it is based on the following two arbitrary assumptions:

1. The claim that two events taking place at two different sites are simultaneous has a content that is independent of the choice of reference system.

2. The spatial distance of the sites at which two simultaneous events take place is independent of the choice of reference system.

Since both the theory of Maxwell-Lorentz and also the principle of relativity are sufficiently supported by experiment, one will be forced to drop the two arbitrary assumptions just mentioned, whose apparent obviousness is based only on the fact that light provides us with *apparently momentary* evidence of events, which occur at distant sites, and that the velocities of the bodies with which we deal in our everyday experiences are small compared to the speed of light c.

By giving up those arbitrary assumptions, the compatibility between the principle of the constancy of the speed of light, which results from Maxwell-Lorentz electrodynamics, and the principle of relativity is achieved. One can hold up the assumption that the same ray of light in vacuum scatters with velocity c not only with respect to the reference system K but also with respect to any reference system K' that is in continuous translational motion with respect to K. One has only to choose appropriately the transformation equations that exist between the space–time coordinates (x, y, z, t) with respect to K and those (x', y', z', t') with respect to K': the system of transformation equations of these four quantities, to which we are led in this way, is called the "Lorentz transformation". This Lorentz transformation should be inserted in place of the corresponding transformation equations, which were considered the only ones imaginable before the formulation of the theory of relativity, which, however, were based on the above assumptions (1) and (2).

The heuristic value of the theory of relativity consists in the fact that it provides a condition to which all systems of equations expressing general natural laws must correspond. Any system of equations must be made to transform into a system of equations of the same form when a Lorentz transformation is applied (covariance with respect to Lorentz transformations). Minkowski gave us a simple mathematical scheme, to which the systems of equations must be able to be reduced in order to behave covariantly with respect to the Lorentz transformations. Thus he achieved the advantage that for the adaptation of the systems of equations to the conditions mentioned it is no longer necessary to actually apply a Lorentz transformation to these systems.

From all this it is clear that the theory of relativity does not at all provide a means to deduce previously unknown natural laws from nothing. It provides only an always applicable criterion that reduces the probabilities. In this sense it is comparable only to the principle of energy or to the second principle of thermodynamics.

The thorough examination of the more general laws of theoretical physics showed that Newtonian mechanics must be modified to correspond to the criteria of the theory of relativity. By examining cathode rays and β-rays (motion of free electric particles) the modified mechanical equations turned out to be correct. In general, the application of the theory of relativity revealed neither a logical contradiction nor a conflict with the results of the experiments.

Here I want to point out only one result of the theory of relativity because it is important for the following exposition. According to Newtonian mechanics the inertia of a system consisting of a set of material points (inertial resistance to the acceleration of the barycenter of the system) is independent of the state of the system. According to the theory of relativity instead the inertia of an isolated system (floating in vacuum) is dependent on the state of the system in such a way that that inertia increases with the energy content of the system. According to the theory of relativity it is therefore ultimately to *energy* that inertia must be attributed. To this, not to the inert mass of material points, indestructibility is to be attributed; the principle of conservation of mass is therefore absorbed in the principle of conservation of energy.

It has been mentioned above that it would be a big mistake if one were to consider the theory of relativity a universal method that allows one to formulate a theory that is surely correct for a field of phenomena empirically investigated as little as one wants. The theory of relativity only massively *reduces* the number of empirical verifications necessary for the formulation of a theory. But there is a field of fundamental importance of which we have so little empirical knowledge that these, connected with the theory of relativity, are not at all sufficient to determine a general theory unequivocally. It is the field of the phenomena of gravitation. The only way to reach this goal is to have the empirical facts accompanied by physical hypotheses to complete the basis of the theory. The following reflections must for now show how we arrive, in my opinion, at the most natural of hypotheses of this kind.

When we talk about the *mass* of a body we associate with this term two definitions logically completely independent one from the other. By *mass*, we mean on the one hand the constant of the body that measures the resistance of the body against the acceleration of the same ("inertial mass"), on the other hand that constant of the body that determines the magnitude of the force that the body undergoes in a gravitational field ("heavy mass"). A priori it is not at all obvious that the inertial mass and the heavy mass must coincide; we are simply *accustomed* to supposing their coincidence. The belief in this coincidence derives from the experience that the acceleration, undergone by various bodies in a gravitational field, is independent of their material. Eötvös showed that the inertial mass and the heavy mass coincide with a very high precision, excluding, by experiments with the torsion balance, the existence of relative deviations between the two masses of the order of magnitude of 10^{-8}.[1]

In radioactive processes enormous masses of energy are liberated in the form of heat, which is released into the surroundings. The disintegration products which are formed during the reaction have, due to the result of the inertia of the energy exhibited above, added together a smaller inertial mass than the raw material of the radioactive disintegration. This change in the inertial mass is for reactions of this type of known heat range of the relative order of magnitude of 10^{-4}. If the heavy mass did not change simultaneously with the inertial mass of the system, the inertial and heavy masses of various elements should be far more different than Eötvös' experiments would admit. Langevin was the first to draw attention to this important point.

---

[1] Eötvös's method is based on the following: gravity and centrifugal force act on a body on the Earth's surface. For the former the heavy mass is determinant, for the latter the inertial mass of the body. If they happen not to coincide, the direction of the resultant of the two (apparent heaviness) would depend on the material of the body. Eötvös proved with his experiments with the high-precision torsion balance the non-existence of such a dependence.

From what has been said one finds with great probability the coincidence between the inertial mass and the heavy mass of a closed (motionless) system; I think that, in the present state of experiments, we should keep to the assumption of this coincidence without any doubt. Doing so we have satisfied one of the most important requirements of physics that must, in my opinion, be demanded of the theory of gravitation.

This requirement implies a broad limitation to the theories of gravitation, as is recognized especially when one combines it with the principle of the inertia of energy. Inertial mass corresponds to every energy, and to every inertial mass heavy mass corresponds; the heavy mass of a closed system must therefore be determined by its energy. To the energy of a closed system the energy of its gravitational field also belongs; this latter energy must therefore contribute not only to the inertial mass of the system but also to the heavy mass.

Gravitational theories have been postulated by Abraham and Mie. Abraham's theory contradicts the principle of relativity, Mie's contradicts the requirement for equality between inertial and heavy mass of closed systems. According to the latter theory, on heating a body its *inertial* mass would *increase* according to the addition of energy, but not so the heavy mass. The latter would even decrease in a gas with increasing temperature.[2]

A theory of gravitation recently postulated by Nordström, on the other hand, corresponds both to the principle of relativity and to the requirement for the gravity of the energy of closed systems, with a limitation indicated below. Abraham's contrary view, set forth in his paper appearing in this journal, is not correct. I believe in general that no valid argument against Nordström's theory can be derived from experiment.

According to Nordström's theory the principle of energy gravity of immobile closed systems is valid as a statistical principle. The heavy mass of a closed (on the whole immobile) system is generally an oscillating quantity whose mean time value is given by the overall energy of the system. The oscillating character of the mass implies that such a system should continuously emit longitudinal gravitational waves. But the energy loss we should expect according to the theory is so low that it should subtract from our perception.

Whoever meticulously studies the theory of Nordström will have to admit that this theory, from the point of view of experiments, perfectly aligns gravitation to the scheme of the theory of relativity (in the narrow sense). If I am however of the opinion that we must not be satisfied with this solution, this has reasons of the gnosiological kind, of which I will speak later on.

---

[2] It is not possible to access these effects experimentally because they are too minute. But it seems to me that there are many reasons to think that the ratio of inertial to heavy mass is guaranteed *in principle*, except for the kind of forms of energies present. According to Mie, one can answer that the equality of inertial and heavy mass is maintained during radioactive transformation only by assuming a special nature of the energy within the atom.

The Theory of Relativity in the Broadest Sense

Classical mechanics as well as the theory of relativity in the restricted sense just discussed suffer from a fundamental deficiency, which no man open to gnosiological arguments can deny. The weaknesses of our physical *Weltbild* that we have to deal with have already been unveiled in all clarity by E. Mach in his extensive investigations into the foundations of Newtonian mechanics, so that what I shall expound here in this connection cannot claim to be new. I want to illustrate the basic point with an example that I have chosen because it is elementary enough to emphasize the essential.

In space two masses stand at a great distance from all other heavenly bodies. They are close enough to each other to exert mutual influence. Let an observer now follow the motion of the two bodies by viewing the starry firmament always in the direction of the line connecting the two masses. He will observe that the line of vision draws a closed line on the visible firmament, which does not change its place with respect to the visible starry firmament. If the observer is endowed with natural reason but has never learned either geometry or mechanics he will conclude, "My masses execute a motion which is causally determined at least in part by the stellar system. The laws, according to which the masses near me move, are also determined by the stars." A man who has passed through the scientific school will smile at the observer's naivety and tell him, "The motion of your masses has nothing to do with the sky of fixed stars; rather it is determined by the laws of mechanics independently of other masses. There is a space R within which these laws apply. These laws are such that your masses remain perpetually in one plane of this space. The stellar system, on the other hand, cannot spin within this space because otherwise it would be torn apart by enormous centrifugal forces. Therefore it is necessarily immobile (at least almost!) if it wishes to exist permanently; from this it follows that the plane on which your masses move always passes through the same stars." Our undaunted observer will reply, however, "Perhaps you are irresistibly learned. But I do not believe in this huge thing you speak of which you call space any more than I believe in ghosts. I can neither see nor imagine anything like it. Or must I imagine your space R as a thin network of bodies to which all other things refer? Then I can imagine next to R also a second net R' moving at will relative to R (e.g. spinning). So your equations hold at the same time in relation to R'?" The educated man will surely deny this. Whereupon the naive man: "But from where do the masses know in relation to which of the "spaces" R, R' etc. they are to move according to your laws, from what do they recognize the space or spaces to which they are to find their bearings?" Now our learned man is in the greatest embarrassment. He will insist that there must be such privileged spaces, but will be unable to point out any reason why these spaces might be distinguished from others. Whereupon the naive man says: "Then until further notice I consider your privileged spaces as a useless invention, and remain of the opinion that the firmament also determines the way in which the masses under examination behave mechanically."

I want to expound also in a second way the infraction against the most elementary postulates of gnosiology, of which our physics is guilty. We will vainly strive to explain what we must mean by the acceleration of a body. We will only succeed in defining the *relative* accelerations of bodies with respect to each other. On the other hand, we base our mechanics on the premise that a force (cause) is necessary to produce an acceleration of a body, pretending not to see that we are not even able to indicate what we mean by "acceleration", precisely because only the *relative* accelerations can be the object of perception.

The danger that lies in our way of proceeding is very beautifully illustrated by a comparison which I owe to my friend Besso. Imagine that you have moved back to a remote time when the Earth's surface was thought to be almost *flat*. Suppose that the following concept reigns among scholars: there is in the world a physically privileged direction, the vertical. All bodies fall in this direction if they are not supported. It may be inferred from this that with respect to this direction the earth's surface is essentially perpendicular and therefore tends towards the plane form. If here the error lies in unfoundedly privileging one direction over the others (fictitious cause) instead of simply considering the Earth as the cause of the fall, in our physics it lies in the unfounded introduction of privileged systems of reference as fictitious causes. Both cases are characterized by the failure to provide a satisfactory reason.

Since not only classical mechanics but also the theory of relativity in the restricted sense exhibit this fundamental shortcoming, I set myself the goal of generalizing the theory of relativity in such a way as to avoid this imperfection. In the first place I realized that general gravitation must be assigned a fundamental role in such a theory. Because from what I have expounded before it follows that every physical process, given that quantities of energy correspond to it, must also produce a gravitational field. On the other hand the fact derived from our experiments that all bodies in a gravitational field fall in an equally rapid way, makes us assume that the processes in a gravitational field take place exactly as relatively to an accelerated reference system (hypothesis of equivalence). Taking this concept as the basis I arrived at the result that the speed of light is not to be considered independent of the gravitational potential. The principle of the constancy of the speed of light is therefore incompatible with the hypothesis of equivalence; it is therefore impossible to reconcile it with the theory of relativity in the restricted sense. This led me to consider the theory of relativity in the restricted sense as exact only for fields where there are no perceptible differences of the gravitational potential. The theory of relativity (in the narrow sense) was to be replaced by a more general theory including the former as a limiting case.

The path leading to this theory can be described only very incompletely in words.[3] The law of motion of a material point in a gravitational field deriving from the equivalence hypothesis can be easily written in such a form as to be completely independent of the choice of definite variables of place and time. Leaving the choice of these variables *a priori* to arbitrariness, thus not favoring any determinate space–time system, avoids the gnosiological objection set forth above. In that law of motion a quantity appears

---

[3] Cf. A. Einstein and M. Grossmann, "Zeitschrift f. Math. & Physik," 62, p. 225, 1914.

$$ds^2 = \sum_{\mu v} g_{\mu v} dx_\mu dx_v$$

which is an invariant, i.e., a quantity independent of the choice of the reference system (i.e., of the choice of the four space–time coordinates). The quantities $g_{\mu v}$, are the functions of $x_1 \ldots x_4$ and are used for the description of the gravitational field.

With the help of the absolute differential calculus, that was developed by Ricci and Levi–Civita on the basis of mathematical research by Christoffel, it is possible, in virtue of the existence of the above invariant, instead of the well-known systems of equations of theoretical physics, to introduce equivalent ones (in the case of the constancy of all the $g_{\mu v}$) that are valid independently of the choice of the space–time coordinates x_v. All such systems of equations contain the quantities $g_{\mu v}$, i.e., quantities that determine the gravitational field. These then influence all physical processes.

Conversely, the physical processes must also determine the gravitational field, i.e., the quantities $g_{\mu v}$. One arrives at the differential equations that determine these quantities through the hypothesis that the laws of conservation of momentum and energy must be valid at the same time for material events and the gravitational field. This hypothesis also subsequently restricts the choice of spatio-temporal variables x, but without awakening the gnosiological objections discussed above. For according to this theory of generalized relativity there are no longer privileged spaces with particular physical qualities. The unfolding of all processes is dominated by the quantities $g_{\mu v}$, which in turn are determined by the physical processes of the whole remaining universe.

This theory fully satisfies the principle of inertia and gravity of energy. The laws of motion of heavy masses are, moreover, such that absolute acceleration (the acceleration with respect to "space") does not seem essential to the appearance of the resistance of inertia, but—as should be required by the reflections above—the acceleration with respect to other bodies.

The theory of relativity in the broadest sense does not mean the abandonment of the previous theory of relativity, but a further development of the latter, which seems to me necessary in view of the gnosiological reasons I have given.

On the Problem of Relativity

Max Abraham

In an article entitled *The new mechanics*, which recently appeared in this journal, I discussed from a historical and critical point of view the two theories of relativity established by Einstein in 1905 and 1913. Since then, the author himself has expounded his views in a very useful article on the psychology of relativity, without, however, refuting the serious objections I had raised against both the "restricted" theory of 1905 and the "general" theory of 1913.

As for the experiments made on cathode rays and β-rays, it is true that these have shown that the inertial mass of electrons is a function of velocity and that, consequently, classical mechanics is at fault on this point. But it has not been possible so far to give these measurements the sensitivity that would be necessary to determine exactly this function. The question whether it is precisely what it should be according to the theory of relativity is still controversial. For this reason it has not been discussed in my article.

The only argument in favor of the theory of "restricted" relativity is the negative result of the experiments made to establish that the motion of the Earth influences electrical and optical phenomena. But when Einstein establishes a parallel between the relativistic criterion deduced from this fact and the two fundamental principles of thermodynamics, it must be observed that such principles have been verified in all the fields of physics and chemistry while serious difficulties oppose the application of the relativistic criterion to non electromagnetic natural forces such as, for instance, gravity. Taking these difficulties into account, it is extremely unlikely that mechanics will be captured by the straight-jacket of Lorentz's group. I do not concern myself here with the question whether most theorists really grant so little importance to these difficulties as to accord their adherence to the definition of space and time described by the 1905 theory of relativity; but I believe, for my part, that in a referendum this

---

M. Abraham, *Sur le problème de la relativité*, "Scientia", 16, 1914, pp. 101–103.

theory would not obtain a majority of the votes except by granting more votes to those who shout the loudest. In any case, Einstein himself would be forced to vote against it, since in the theory of relativity of 1913 he abandons the postulate of the constancy of the speed of light and, consequently, the ideas about space and time set forth in the theory of 1905.

Between the theories of gravitation of G. Nordström and of G. Mie, Einstein pronounces in favor of the former established under his influence; he reproaches Mie's theory for not satisfying the postulate of the identity between gravitational mass and inertial mass. Well Mie has shown[1] that the theories of Nordström and Einstein are far from rigorously satisfying this postulate. He showed how in these theories the weight is proportional to the energy only for isolated systems that are in equilibrium or in stationary motion and that it is so only by virtue of some secondary hypotheses, which on the other hand contradict themselves. Under these conditions, it seems that Einstein has no right to assert as an objection to Mie's theory the fact that, in this theory, the temperature exerts on the weight/energy ratio an influence that is not experimentally appreciable. This objection is justified only when the rigorous validity of the weight-energy principle is postulated for any system. Restricting the validity of this principle, as Einstein and Nordström do, to the case of systems that are in true or static equilibrium, we must consider as admissible all the theories that satisfy this principle within the limits of experimental errors.

In my opinion, however, if the theory of relativity establishes, recalling Michelson's experiment, the general and rigorous validity of the postulate of relativity, and not only its validity within the limits of experimental errors, we have in the same way the right to make the rigorous validity of the law of the conservation of weight in static gravitational fields a fundamental principle of mechanics. Yet no relativistic theory of gravitation satisfies this postulate. It is precisely because I have recognized that the postulates of relativity and the conservation of weight are irreconcilable that I have been led to base the theory of gravitation on the principle of the weight of energy, making it independent of the postulate of relativity.

And here I come to the imaginary dialogue established by Einstein between a "simple" man and an educated man. If neither of the two interlocutors tells us anything new, it is nevertheless true that this discussion is charming and well suited to bring out the crux of the matter. But when the simple man says, by way of conclusion "Then I consider your privileged spaces as a useless invention, and remain of the opinion that the firmament also determines the way in which the masses under examination behave mechanically," the educated man cannot be satisfied with this consideration. The author of the "extended" theory of relativity would have to show, if he did not want to be satisfied with the approval of simple men, that in this theory the fixed stars really exert, in spite of their distance, a perceptible influence, and that the centrifugal forces are here caused by a rotation relative to them and not by an "absolute" rotation. But even in Einstein's theory the contribution of the fixed stars to the gravitational field is too weak to permit such a formulation. At the beginning

---

[1] G. Mie, *Remarques sur la théorie de la gravitation d'Einstein*, "Physikalische Zeitschrift", 15, p. 115 and 169, 1914.

Einstein probably allowed himself to be guided, basing himself on his "equivalence hypothesis", by the hope of arriving at a general theory of relativity that included rotational and accelerated motions. He should, however, realize by now that he has in no way achieved this result and that the balance of the theory of relativity presents here a gap to be filled.

Can Gravitation Be Explained?

Arthur Stanley Eddington

In Einstein's theory the phenomenon of gravitation is shown to be a necessary consequence of the "curvature" (or non-Euclidean geometry) of the space and time around heavy masses. It is sometimes said that Einstein *explains* gravitation by attributing curvature to space and time; but this scarcely conveys the right impression of the purpose and achievement of his theory. The curvature of space–time is intended to be a description rather than an explanation. A region where gravitational effects are perceived differs in some way from a region free from gravitation; the intrinsic nature of this difference is necessarily beyond our comprehension, and we can only describe it by symbolism or analogy. Newton described the difference as consisting in the existence of an agent "absolute force", *i.e.,* something acting in a way analogous to the force of muscular effort with which we are directly familiar. Einstein describes the difference as a curvature, and his symbol takes shape in our minds by analogy with the curvature of surfaces. We believe that by Einstein's symbol a more accurate quantitative description of the facts is possible; but it does not profess to indicate the mechanism of the phenomena discovered and described in this way.

Nevertheless Einstein's theory makes some advance towards an explanation; at least it shows us more definitely what it is that requires explanation. Accepting his standpoint we have realised for the first time that a world without gravitation (without curvature) would be more specialised and stand more in need of explanation than the actual world disturbed by gravitation; just as the occurrence of a level lawn requires more explanation than the occurrence of an undulating field. This is an inversion of former ideas. The beginner in mechanics is taught that for every particle to move with uniform velocity in a straight line is a natural condition, but that any deviation from this orderly state must be explained as due to some outside cause. Surely this is to begin at the wrong end. *Order* requires explanation; *disorder* requires none—it is

<hr>

A. S. Eddington, *Can Gravitation be Explained?*, "Scientia", 33, 1923, pp. 313–324.

merely the indication of absence of controlling mechanism. It is because gravitation, though perhaps first appearing to us as a deviation from a more ideal type of order, does nevertheless impose an order on the motions, that Einstein's theory leaves something to be explained. Not the deviations, but the law of the deviations—not the existence of gravitation, but the law of gravitation—is the mystery to be explained. It is strange how we have puzzled ourselves by setting up a negation as an entity, and then tried to explain the occurrence of this entity. But having appreciated this, we still have the task of explaining the positive aspect of gravitation contained in the law which it recognises.

Our problem then is to explain the occurrence of the order observed in motions of the planets and other bodies, this order being expressed by the law of gravitation. Einstein has traced it so far as to recognise that it is reducible to a law governing the curvature of space and time. Can we see any likely reason why the curvature of space and time should be subject to a law of any kind? Or can we suggest any mechanism by which the properties of space–time are adjusted so as to conform to this law? The first difficulty is that we seem here to be approaching the extreme limits of physical explanation. A particular phenomenon is said to have been explained when it has been traced to its source in the great fundamental laws of dynamics and physics; but here the problem before us is whether one of those fundamental laws can itself be explained. In pursuit of explanation must we not ultimately arrive at some statement of fact about the way the world is built which does not admit of further simplification? The final substratum or æther cannot be devoid of properties; can we do more than discover those properties without seeking to account for them? Nor ought we to expect that these properties will be of a type familiar in our experience of material bodies—rigidity, inertia, permanence for whatever may be the nature of the substratum of the world, it is at least different from gross matter. If then Einstein's law of curvature of the world is merely a statement of the essential nature and properties of that *continuum* from which all phenomena take their origin, we can scarcely hope to reach any further explanation. This stage is usually supposed to have been reached in the theory of the electromagnetic field; the phenomena are reduced to a set of equations given by Maxwell and these are regarded as defining the æther in a way unlikely to be further simplified. We could only account for the properties of the æther by regarding it as a structure composed of some sub-æther not possessing those properties intrinsically; but the sub-æther itself could not be devoid of all properties, and these would have to be explained by reference to a sub-sub-æther—and so on *ad infinitum.*

Although we might have feared that the search for an explanation of the law of gravitation would be barred in this way, it appears that in fast it is not so barred. The law of world-curvature is not the expression of a property of the underlying substratum. It arises in a much more subtle way through the methods employed in investigating physical phenomena and the manner in which we represent to ourselves the results of those investigations. We shall see that from this point of view the actual law is one which might have been anticipated as inevitable.

Einstein's law of gravitation (or world-curvature) is usually expressed as *a* series of ten equations held in empty space, which are written in the mysterious symbolic form

$$G_{\mu\nu} = \lambda g_{\mu\nu}$$

The reader must turn elsewhere to learn how from these equations Einstein deduces that bodies will attract one another approximately according to Newton's law of the inverse-square, but with minute deviations verified observationally in the motion of the planet Mercury and the deflection of light-rays near the sun. Our task here is to understand why these particular equations and no others are satisfied in nature. One point, however, should be noticed. Newton's law of gravitation determined the force of attraction on a planet, but it had to be supplemented by other laws (the laws of motion) to discover how the planet would move under the influence of that attraction. Einstein's law needs no supplementing; it contains within itself the laws of mechanics—the conservation of energy and momentum—and determines both the field of gravitation and the motion of a particle in that field.

The law is frequently seen stated as $G_{\mu\nu} = 0$; that is to say, the constant λ in the form given above is considered to be zero. The term $\lambda g_{\mu\nu}$ is usually called the "cosmical term" in the law of gravitation, because it is too small to be appreciable in the ordinary practical problems relating to the solar system, and could only rise to significance in problems on the vastest scale embracing the stellar universe and regions beyond. This term is, however, of great importance for our explanation of gravitation and it may be of interest to trace its history. Einstein first stated the law without the cosmical term, and introduced the term later. It has the effect of making the whole volume of space finite (although without a boundary). The non-mathematical reader will probably find it equally difficult to conceive space as infinite and space as finite—a well-known dilemma; but the mathematician can quite well grasp the conception of unbounded finite space, whereas infinite space is beyond the conception of everybody. Thus a quite unnecessary difficulty about contemplating infinity was removed, and in the view of the writer this is the primary recommendation of the cosmical term. The term was also supposed to remove a difficulty in regard to absolute rotation; but it may be doubted whether the difficulty was a real one, and whether the change cleared it away. Later on Weyl in working out his more general theory arrived directly at the law of gravitation with the cosmical term included; the constant λ performs an important function in his theory and cannot be put equal to zero, although it is certainly very small. We can scarcely do better than follow Einstein and Weyl in accepting the cosmical term as a necessary part of the law.

When written out in full the equations are of extreme complexity and forbidding technicality. Is there no more concise way of stating just what Einstein's law amounts to? Yes; by introducing geometrical nomenclature we can compress the law into a statement which is certainly not long-winded, and even seems to be on the verge of intelligibility! It runs.

The radius of spherical curvature of a three-dimensional section of the world taken at any point and in any direction is always the same constant length.

This is exactly equivalent to the equations $G_{\mu\nu} = \lambda g_{\mu\nu}$, the constant length being $\sqrt{(3/\lambda)}$.

It seems as though with a little effort we might be able to picture the geometrical meaning of that statement, but I am afraid the pope is illusory. The principal difficulty is that no one is able to picture the curvature of a four-dimensional continuum because there is nothing in our experience which at all resembles it. Many geometrical conceptions based on ordinary space can be extended without much effort of the imagination to space, of higher dimensions; but curvature is more baffling. Our experience of curvature of surfaces gives us simply no conception of the variety of contortions possible to a continuum of four dimensions. For one thing the continuum must be provided with six extra dimensions (ten dimensions in all) to twist about in, before it exhausts the variety of curvatures of which it is capable. An ordinary surface with radii of curvature the same at all points and in all directions would necessarily have the dull uniformity of a sphere. Not so a four-dimensional continuum; the law stated· above imposes homogeneity and isotropy for a certain characteristic, but nothing like complete uniformity.

It is perhaps desirable here to emphasize the fact that in speaking of the curvature of space and time Einstein is using a symbol or analogy which is not to be understood too literally. Space and time together contain four dimensions, and if the combination were actually bent it would have to be bent in the direction of some fifth dimension outside both space and time. We do not believe that for a moment. The significance is that, owing to its non-Euclidean geometry, space–time has certain metrical properties, which if we were restricted to Euclidean geometry (which we are not) could only be accounted for by assuming it to be curved; we reject the restriction as unnecessary and irrational, but at the same time we find it convenient to retain the nomenclature which would result if the restriction were in force.

It would be a waste of time trying to visualise the curvature referred to in the geometrical statement of Einstein's law. What is essential to our purpose is that to every point and to every direction in empty space–time a significant length[1] can be associated; and whereas in an arbitrary continuum this length would vary from point to point and from direction to direction, Einstein's law asserts that in actual space–time this length is a universal constant. The remarkable analysis of the relativity theory shows that the motion of the planets, in their appointed orbits and all the orderly phenomena of gravitation result inevitably from this distinguishing characteristic of actual space–time; and accordingly our problem is to find out why actual space–time should be of this particularly limited type. We shall call this important length the radius of curvature in the direction considered, because it will be identifiable to mathematicians as the radius of spherical curvature of the corresponding section; but we do not ourselves need to trouble about 'that identification. Although the length is not a distance between two particular points it may be thought of as intrinsic in and characteristic of the region, much as the free period of a mechanical vibration

---

[1] We are dealing with time as well as space; but for simplicity I here confine attention to space, since that sufficiently illustrates the nature of the argument.

may be thought of as a time-interval intrinsic in an object even if no oscillations are actually occurring.

Absolute size is as unimaginable as absolute velocity. A uniform change of scale of ourselves and all our surroundings would be undetectable if carried out with entire thoroughness. Size has a definite meaning for us, but it is a relative meaning; and it is necessary to examine precisely what this meaning is. Size is determined by reference to material standards, and we must not imagine that there can be any definition of size which dispenses with this reference to material objects. We say that a length AB is one metre if the standard metre (a material rod), when brought into the required position and orientation coincides with AB. There are also indirect (*e.g.* optical) methods of determining the length AB; but there are justified because it is found by experiment that they give the same result as actual transference of the standard metre. Primarily therefore the measurement of a length involves transference of the material standard to the position and orientation required, followed by a direct comparison. No alternative method can be accepted unless it has been proved to be equivalent to this. Our previous statement of the law of gravitation can thus be amplified as follows. If the standard metre-rod is transferred to a particular point and direction, the radius of curvature at that point and in that direction will be a certain constant numerical multiple of the extension of the rod.

But this statement seems to contain an inversion of ideas "putting the cart before the horse". Would it not be more natural to say that when the standard metre is transferred to a new point and direction it takes up an extension which is a constant numerical submultiple of the corresponding radius of curvature there? There is, of course, no essential difference between the two statements; if one is true the other must be true; but the implication suggested is widely different. The law of gravitation no longer hints at a mechanism of adjustment of curvature in void regions; it is a statement of the adjustment of the extension of a material object as it is moved about.

When the standard metre is moved to a new position, how does it determine the extent of the region it ought now to fill? How do its individual molecules decide on their average spacing and settle down to the positions which determine the extension of the rod? Our first impulse is to reply that the rod solves the problem by remaining of the same size as it was before. But that is to argue in a circle. There is no absolute size. The rod is our standard metre and however it behaves it can never be anything but one metre long. To say that it will continue to extend one metre in the new position is merely to say that it will have to be the same size as itself. The fact that we are conventionally going to call its length one metre, and to measure all other lengths at the same point and in the same direction by comparison with it, cannot help the rod to solve the problem how to arrange its molecules in the new circumstances.

The spacing of the molecules will not be determined by memory of a former state but by the interplay of the various conditions which exist in the locality and affect the equilibrium. The rod has to find out its proper extension x having regard to these conditions of equilibrium; the mathematician might formulate the same problem as a series of equations, and solve them to find x. At present we are too ignorant to know what are the proper equations to write down; let us however anticipate future knowledge, and suppose that the equations have been discovered and solved, and that the mathematician has arrived at his answer

$$x = \text{something or other.}$$

It is well-known that in all such equations the physical dimensions of both sides must be the same. Since x is a length, the other side of the equation must also be a length. Moreover the problem was to find the length of the rod as determined by the conditions of the region in which it is placed, so that the length on the right side of the equation must be characteristic of the conditions of the region. Thus whatever may be the unknown equations, their solution will necessarily be of the form

$$\text{Length of rod} = \text{pure number} \times \text{a length intrinsic}$$
$$\text{in the region before the rod was inserted.}$$

Since the problem must be solved afresh for every position and direction of the rod, the length intrinsic in the region must also be associated with position and direction.

It is not easy to find in an empty region of space–time a characteristic length associated with position and direction. But we have already noticed one such length, viz. the radius of curvature above-mentioned. Probably other lengths could be constructed, but these are highly complicated and it is difficult to regard them as more than ingenious exercises of the mathematician. Unless the conditions of adjustment of material structures are far more complicated than we have reason to expect, it is inevitable that the solution should take the form

$$\text{Length of rod} = \text{pure number} \times \text{radius of curvature}$$
$$\text{(for the position and direction considered)}$$
$$\text{in the region where the rod was inserted.}$$

But this is precisely what Einstein's law asserts.—Our last statement of the law was that "when the standard metre is transferred to a new point and direction, it takes up an extension which is a constant numerical submultiple of the corresponding radius of curvature there".

Instead of demanding a very specialised structure of the world we see that the law is so inevitable that we have arrived at it as the necessary solution without yet knowing the precise equations to be solved. Instances are not infrequent in physics, where by an argument from dimensions, the solution of a problem can be found (apart from an unknown numerical factor) without any detailed study of the equations; we survey the various factors which can affect the solution, and if these can only be grouped in one way to give the right physical dimensions, this must necessarily be the solution (apart from an undetermined multiplier). The present argument is of this type.

The radius of curvature is the same at all points and in all directions simply because we accept it as the natural standard of length. It is too transcendental and inaccessible to be used directly as the practical standard; but any material structure adjusts its extension to a definite fraction of this. natural standard (having nothing else to compare itself with), and so can be used equivalently as a more practical standard. In place of a separate standard at every point and in every direction, we can now use a single material rod as providing a standard for everywhere, since the material object can be moved about, whereas we cannot transfer the radius of curvature even in imagination. Since all the observations and deductions of physics are based on the acceptance of a material structure as standard of length, the resulting laws of physics will conform to the original convention that the radius of curvature is the standard to be regarded as constant. The law of gravitation is nothing more than a reiteration of this convention.

To sum up, the explanation of gravitation is divided into four stages:

1. We must realise that it is only in so far as gravitation is a law that the phenomena require explanation. The Newtonian traditions tends to emphasize what may be described as the *negative* aspect of gravitation—the substitution of less simple order for more simple order. We must dwell rather on the positive aspect and the mystery why there is any order at all in the world; and we seek an explanation of that constraint of phenomena which is manifested in its most varied form in the neighbourhood of heavy bodies, and comprises as a particular case the limiting condition of uniform motion in a straight line in regions remote from heavy masses.

2. This order is found by Einstein to arise from the fact that the equations $G_{\mu v} = \lambda g_{\mu v} G_{\mu v} = \lambda g_{\mu v}$ are satisfied in empty space. The deduction is long and mathematical, and constitutes great scientific achievement. We do not discuss it here since it is the well-known theory on which much has been written; there has been no lack of attempts to explain it in non-technical language so far as it admits of such treatment.

3. The interpretation of the equations $G_{\mu v} = \lambda g_{\mu v}$ is that space–time has a homogeneity and isotropy as regards a certain characteristic. The order, which we are seeking to explain, arises because actual space–time is not the most general type of metrical continuum imaginable by a mathematician, but is limited in this way. The property which is homogeneous and isotropic is a curvature (defined in a certain way); but it turns out to be unnecessary to our argument to enter on the question of its exact definition. The task before us is to explain why there is this kind of isotropy and homogeneity in the world.

4. The mystery of this isotropy and homogeneity disappears when we realise that it is not intrinsic in the external world, but in the measurements which we make of that world. Every measurement which we make is a comparison—a comparison of the region surveyed with the surveying apparatus. We must not ignore the second partner in the comparison, and suppose that every regularity in the results of our survey springs from a regularity intrinsic in the region surveyed. It may be a regularity of adjustment of one partner to the other. We have seen that from

this point of view there is a perfectly natural explanation of the homogeneity and isotropy contained in our measurements. The surveying-apparatus has no *absolute* size and can only take up a definite extension by adjusting itself in a constant ratio to the linear characteristics of the region surveyed; when therefore we attempt to measure those linear characteristics with the apparatus we merely reproduce the constant ratio.

It is a well known illustration of the special theory of relativity for uniform velocity, that on a planet travelling with high speed objects expand and contract as they change their orientation according to the Fitzgerald-Lorentz law; and these changes are from their very nature undetectable by any measurements made on the planet. In just the same way if deviations from the above-mentioned isotropy and homogeneity of curvature occurred in any region they would be undetectable by any physical measurements; the same deviations affect the measuring appliances as well as that which is measured and are eliminated in the comparison. We fail to find any deviations and consequently we announce that our survey of the world indicates complete homogeneity and isotropy of curvature, and thereby claim to have discovered a law of nature $G_{\mu\nu} = \lambda g_{\mu\nu}$. But the fact is that deviation from this law is unthinkable when we reflect on the nature of the operation we are performing when we make the measurements in question.

The space and time with which physics concerns itself—actual space–time—is an abstraction of the extensional relations of material objects. General space, with which the pure mathematician concerns himself, is a product of the imagination having no reference to material things. When we compare actual space–time with the mathematician's general space, we notice that the former is more specialised; and ultimately we trace the orderly phenomena of gravitation to this specialisation. Our first impression is that the specialisation is indicative of some controlling mechanism or design in external nature; but the true explanation is probably simpler. Why should our construct of actual space, abstracted out of the extensional relations, be judged by comparison with the mathematician's space constructed on different principles? The comparison is irrelevant. Let us instead turn to examine more minutely that process of abstraction by which we have chosen to construct actual space out of material extension; we then realise that it is not arbitrarily specialised but is of as general a character as could conceivably be consistent with such a mode of construction.

To give another illustration of the argument, consider a normal hydrogen atom which consists of one proton (or positive nucleus) and one electron at a distance apart which may well be used as the fundamental standard of length since it is accepted as a constant of nature. Keeping the proton at a fixed point, orientate the atom in all possible directions; the electron will then describe a certain locus. The atom is merely the mechanical means of making evident these loci intrinsic in space–time around every point. The loci have no shape or size in any absolute sense; but we adopt them as the standard of symmetry and equality, to which the relative terms "shape" and "size" are referred. Spheres are conventionally drawn by orientating a standard of length in all directions, and the locus is called a sphere because of its method of construction—not from any quality discovered in the locus after its construction. In this way we have natural loci termed equal and spherical strewn all over space–time; when therefore we study the geometry of space–time we ought everywhere to come across these equal and spherical loci. That is just what we do find, and the discovery is called the law of gravitation.

Masses in the Theory of Relativity

Enrico Fermi

The great conceptual importance of the theory of relativity, as a contribution to a deeper understanding of the relations between space and time, and the lively and often impassioned discussions to which it has consequently given rise even outside strictly scientific circles, have perhaps somewhat distracted attention from another of its results which, though less dramatic and, let us admit it, less paradoxical, has nevertheless consequences no less noteworthy in physics, and whose interest is probably destined to grow in the subsequent development of science.

The result to which we allude is the discovery of the relationship between the mass of a body and its energy. The mass of a body, says the theory of relativity, is equal to its total energy divided by the square of the speed of light. Even a superficial examination shows us that, at least for physics as it is observed in laboratories, the importance of this relationship between mass and energy is such as to obscure considerably the importance of other consequences which are quantitatively very small but to which the mind becomes used with more effort. Let us take an example: a body one meter long, moving with the speed, respectable enough, of 30 km per second (roughly equal to the speed of the Earth's motion through space) would always appear one meter long to an observer dragged along by its motion, while to a stationary observer it would appear one meter long minus five millionths of a millimeter. As you can see, the result, strange and paradoxical as it may seem, is nevertheless very small, and it is to be assumed that the two observers will not start to argue over so little. The relationship between mass and energy on the other hand certainly brings us to some impressive figures. For example, if we succeeded in freeing the energy contained in

---

E. Fermi, *Le masse nella teoria della relatività*, in R. Contu, T. Bembo (Editors), *Augusto Kopff. I fondamenti della relatività einsteiniana. Valore e interpretazione della teoria negli scritti originali di A. Aliotta, E. Bianchi, G. Boccardi, A. Bonucci, P. Burgatti, V. Cerulli, P. Emanuelli, F. Enriques, E. Fermi, G. Fubini, G. Gianfranceschi, M. La Rosa, Q. Majorana, E. Rignano, U. Spirito, A. Tilgher, E. Troilo, É. Borel, H. Weyl*, Milan, Hoepli, 1923, pp. 342–344.

S. Linguerri and R. Simili, *Albert Einstein—Italian Memories*, History of Physics,
https://doi.org/10.1007/978-3-031-52950-4_14

a gram of matter, we would obtain an energy greater than that developed in three years of uninterrupted work by an engine of one thousand horsepower (comments are useless!). It will be rightly said that it does not seem possible that, at least in the near future, a way will be found to set free these frightful quantities of energy. On the other hand this is something we can only hope for because the explosion of such a frightful quantity of energy would have as its first effect that the physicist who had the misfortune to find the way to produce it would be reduced to smithereens.

But even if such a complete explosion of matter does not appear possible for the moment, for some years experiments aimed at obtaining the transformation of chemical elements one into the other have already been in progress. This transformation, which occurs naturally in radioactive bodies, has recently also been obtained artificially by Rutherford who, by bombarding some atoms with α-particles (corpuscles launched with great velocity by radioactive substances), has succeeded in obtaining their decomposition. Now some energetic exchanges that the relationship between mass and energy allows us to study very clearly are connected to these transformations of the elements one into the other. To illustrate this we may use a numerical example. We have reason to believe that the nucleus of the helium atom is made up of four nuclei of the hydrogen atom. Now the atomic weight of helium is 4.002 while that of hydrogen is 1.0077. The difference between the quadruple of the mass of hydrogen and the mass of helium is therefore due to the energy of the bonds that unite the four hydrogen nuclei to form the helium nucleus. This difference is 0.029 corresponding, according to the relativistic relationship between mass and energy, to an energy of about six billion calories per gram-atom of helium. These figures show us that the energy of the nuclear bonds is a few million times greater than that of the most energetic chemical bonds and explain to us how against the problem of the transformation of matter, the dream of alchemists, the efforts of the most able minds have been broken for so many centuries, and how only now, using the most energetic means at our disposal, we have succeeded in obtaining this transformation; in quantities so small as to escape the most delicate analysis.

These brief hints suffice to show how the theory of relativity, besides giving us a clear interpretation of the relations between space and time, will be, perhaps in a near future, destined to be the keystone for the resolution of the problem of the structure of matter, the last and most arduous problem of physics.

The Relativistic Theory of Time

Hans Reichenbach

While its physical elaboration appeared from the outset to be complete, the philosophical basis of the theory of relativity was revealed to scientists only gradually. It is no doubt typical of Einstein to base his work on a clear vision of fundamental philosophical concepts, but the creator of this great theory owes this vision to an intuition rather than to the elaboration of a gnosiological theory. On the other hand, the creator of this great theory concentrates first on *physical* elaboration, and if he were to devote more time to determining the *philosophical* basis of his theory, this would only harm his studies. The extreme brevity of his observations on this subject has caused a certain number of his opponents to reproach him with a lack of philosophical understanding and to presume to assign to his work a purely mathematical meaning—a misunderstanding which seems ridiculous to those who know the physical theory and above all Einstein's mental form. When a physicist, in order to solve difficulties of a physical order, refers so brilliantly to the theory of knowledge, can it be said that he is incompetent in philosophy? Rather, I think it is worth modestly trying to find out, through philosophical studies, what he has done to the benefit of this discipline, and to grant him the right to construct his new physics, without continuing to concern himself with the results of philosophy.

In the history of philosophy it has happened more than once that the natural sciences have provided many more ideas than the philosophy of the schools. The overcoming of scholasticism is due to exact science, to a Copernicus, a Kepler, a Galileo, without whom Descartes and Leibniz, themselves partly mathematicians, could not have founded their theories of knowledge. Are we going to witness again, in our era, a turning point in the philosophical field? Will mathematics and physics advance on the path of knowledge better than those who have been established as judges on the subject?

H. Reichenbach, *Die Relativistische Zeitlehre*, "Scientia", 36, 1924, pp. 361–374.

S. Linguerri and R. Simili, *Albert Einstein—Italian Memories*, History of Physics,
https://doi.org/10.1007/978-3-031-52950-4_15

We do not want *to judge*, but first let us try *to understand*. What is the philosophical meaning of Einstein's discoveries? We will only be able to answer this question if we apply to our philosophical method the same rigor which characterizes the mathematician. We will limit ourselves to examining a particular problem instead of pronouncing on everything with statements of empty superficiality. We will only investigate the meaning of the first novelty introduced by Einstein, namely the *relativity of time*.

I

Let us begin at once with the need to rectify most expositions of the theory of relativity for the general public. In these writings we usually read: "What appears simultaneous to one observer, appears non-simultaneous to another observer moving towards the first one". It is an attempt to represent the relativity of *time* as a relativity of the *observers*. Apparently it is believed that in this case we are dealing with differences that would be comparable to the diversity of spatial perspective of two observers far from each other and would have their foundation in the subjectivity of perception. Well, this means that the logic on which Einstein's theory of time is built is not well understood. This only considers the *logical* conditions of knowledge and not the *psychological* ones. It is not at all concerned with the perception of the observer, but with the nature of the object to be known. It discovers the presuppositions of knowledge and not of cognition as a psychological act.

Consider in fact the perception of the observer: he *cannot* in any way tell us anything about the simultaneity in different places that Einstein's theory deals with. The observer is fixed in his position in space: he only receives information, signals from distant points and can only *immediately perceive* the simultaneity of their arrival in the place where he is. Of course, this place is not strictly a mathematical point, but it can be considered as practically infinitesimal compared to the distances that light travels in a few seconds and to which the theory of relativity relates. We can consider the arrival of a signal to the observer as a *coincidence*, an isolated event. We will therefore adopt, without variations, the way of considering *simultaneity in the same place* by ancient physics. The *logical* problem, which arises beyond perception, is the following: how does the observer come to *order in time events distant from each other in space*?

The first answer will be: "through physical measurements". The observer measures the spatial distance and divides it by the speed of the signal. He thus obtains the time in which the path was traveled. If a luminous ray from Sirius arrives on Earth at the same time as a luminous ray emitted by the Sun, it is possible to calculate, on the basis of the distances of these stars and the speed of light, at what instant each of these rays departed; thus obtaining the comparison between the time of the Sun and that of Sirius.

This is certainly true. But to operate in this way you need to know the speed of light. How can we measure it?

There is *only one* method to measure the speed of a signal and we can, schematically, expound it in the following way. Let us imagine two clocks located at two points distant from each other. The signal starts from the first point, let us say at 12:00. It arrives at the second point at 12:05. It has therefore taken 5 min to cover the distance that I now measure; with a division we obtain its velocity. This is the only possible method to measure a velocity.

But is that really true? Did not Fizeau measure the speed of light in an entirely different way? He sent a ray of light to a distant point; he made it be reflected there and return. Fizeau had to measure time only at the point of departure, without worrying about the instant when the light beam struck the mirror. But in this way he obtained only the *sum of* the times of the outward and return *journey*; he could not measure what interests us, that is, the velocity *on a one-way journey*. Our statement is therefore correct.

We note, however, that our way of measuring velocity leads to a difficulty. To measure velocity, we need *two* clocks placed in different places. In order for the difference in time that we read on the clocks to make sense, we must of course *compare them*, that is, we must check whether they indicate the same figures *at the same time*. We wanted to measure the velocity for the sole purpose of obtaining a way of establishing simultaneity in two places far apart. And so we fall into a vicious circle: in order to establish the simultaneity of events far apart, we must know a velocity, and in order to measure a velocity, we must be able to assess the simultaneity of events far apart.

Einstein pointed out the solution to escape from this vicious circle: it is not possible *to recognize* the *simultaneity* of distant events; we can only *define* such simultaneity. This simultaneity is *arbitrary*; we can introduce any way of defining it without errors arising from it. Since, when we take measurements, we will always obtain as a result the simultaneity previously introduced by definition, this procedure can never lead to a contradiction.

The following objection will probably be raised: is this Einstein's theory? In the theory of special relativity, does he not give a rule as to how to arrange the figures on the clocks? He lays them down in such a way that the outward and return velocities of light become equal. Yet it would be a mistake to believe that this rule has any claim to be *truer* than any other. It leads only to simple relationships between numbers; but this is true only when a certain objective condition is respected, that is, when there is no gravitational field. Otherwise, even the simplicity of this particular definition is lost. But another, more complicated definition of time is equally admissible even in a space in the absence of a gravitational field. It leads to a quantitative and univocal description of all natural phenomena and this is sufficient to assign to it the character of *truth*. It is precisely in the theory of general relativity that Einstein succeeded in demonstrating that the choice of time is totally arbitrary: according to this theory, every kind of definition of time is legitimate. Therefore only Einstein's general theory is the true solution to the problem of the gnosiological theory of simultaneity.[1]

---

[1] The statement of the theory of general relativity goes even further, from the physical point of view, than the relativity of time considered from the point of view of the theory of knowledge.

It is now clear that for Einstein it is therefore the *logical* problem of temporal order and not the *psychological* problem. The question of the definition of time exists equally for each observer and already the individual has at his disposal all the possibilities of definition. Distributing two different definitions of time between two observers is only a better *illustration* of things. But the observer "at rest" can define time in the same way as the observer "in motion"; at this point, for his system, the speed of light is not the same in both directions. This is not a difference in viewpoints, but a difference in the logical assumptions of measurement; these assumptions must be established by an arbitrary definition before a measurement can be made.

II

Only starting from this point of view can we address the gnosiological questions that relate back to the theory of relativity of time. We must clearly expose the problem from a logical point of view before we can answer this other question: is Einstein's solution *admissible*?

We agree that it is impossible to measure simultaneity; but then it will be objected that no conclusion can be reached as to the true state of things. In many cases physics reaches the limit of its possibilities as far as measurement is concerned. Are we then to draw the conclusion that the quantity to be measured does not exist? Similarly, it is not possible to determine the exact number of molecules contained in a cubic centimeter of air, and it may be said with absolute certainty that it will never be possible to count each of these molecules. But are we therefore authorized to conclude that such a number does not exist? On the contrary we will state that there always exists an integer that exactly defines this quantity. Einstein's error would then be to have confused the *impossibility of measurement* with *objective indeterminacy*.

Those who raise this objection fail to notice a fundamental difference. There is an impossibility of measurement due to the fact that our technical means are limited. I would call it a *technical impossibility*. There is also an impossibility of *principle* which is based on logical reasoning. Even if we were in possession of a perfect experimental technique we would not escape this impossibility of principle. Thus there is an impossibility of principle relating, for example, to the fact that we will never discover whether the example meter kept in Paris is really a meter. We can infinitely perfect our geodetic instruments but they will never be able to give us information about it, for the simple fact that it is not possible to establish absolutely what a meter is. So we consider the example in Paris the *definition* of a meter. It has been arbitrarily decided to take it as a unit of measurement, and wondering whether it really represents such a unit is now meaningless. The same happens

---

See my detailed account in *La signification philosophique de la théorie de la relativité*, in "Revue philosophique", July 1922. In this paper I have also considered another definition of simultaneity other than that based on signals. A very detailed research on these matters can be found in my *Axiomatik* recalled below, § 2–3 and § 21–28.

with simultaneity: here it is not a question of technical limitations, but of logical impossibility. The presumption to measure simultaneity, without having first fixed a rule to constitute it, has no more sense than measuring a length in meters without having previously adopted a given length as a meter. The comparison which is made with the calculation of the number of molecules is erroneous because this number is precisely defined, but it is in this case a technical impossibility. This number which technique does not allow us to calculate, we can at least determine by *approximation* with increasing exactness, but even this is not possible when we are faced with logical impossibility. We will not be able to determine even *approximately* the simultaneity of events distant from each other until a rule exists. However, even assuming this, the technical impossibility will continue to exist. Einstein's simultaneity itself can be practically defined only approximately.

So there is no reason to consider this objection. Relativity of time means that there is something arbitrary in the logical basis of our measurement which has nothing to do with limits of a technical character. It means that the property of simultaneity is not given in objective events but enters into the description of nature only by virtue of our mental form. It means that we must by definition establish an order of natural events in the four coordinates of space and time before we can apply it. Here is a concept of Einstein's to which one can oppose no argument.

III

Here, however, another objection arises. If we agree that it is in principle impossible to measure the simultaneity of distant events and that we must limit ourselves to giving a definition, will this definition be *arbitrary*? Will it not, on the contrary, be subject to certain restrictions that reside in the nature of our *mind*? It is acknowledged that there are presuppositions of the knowledge of nature which belong to the mind. But are there not for such presuppositions restrictions of a particular kind which arise from the fact that, since we are beings endowed with reason, we must grasp the world of phenomena intuitively?[2]

This is the objection raised by all those who accept Einstein's theory of time on an intellectual level, but who cannot grasp it intuitively. "I cannot picture it to myself", is the answer that comes even from great thinkers. It seems that there is a psychic limitation that leads one to refer to the absolute meaning of simultaneity. Let us analyze it.

Let us take two events: I knock on the wall and you knock on the wall on the other side. Did the two events occur simultaneously? No, I distinctly heard you knock after me. Einstein says these events could be said to be simultaneous? That is impossible.

---

[2] This is also the concept of Kantian philosophy. A detailed exposition of these questions and analysis of this philosophical conception can be found in my paper *Relativitätstheorie und Erkenntnis* a priori, Jul. Springer, Berlin 1920. I set forth other philosophical conceptions of the theory of relativity in "Logos", X, Tübingen 1922, p. 316.

This is where intuition fails. At this point, limits of another kind impose themselves on my mind.

So what do I mean when I say that the two events are not simultaneous? I simply assert that my *perception* of these events possesses a quality which I call "non-simultaneous". But in order to go from perception to events, a *succession of syllogisms* is required. When I examine the psychic limitation more closely, I find that it exists only for *perception* and that I have transferred it to *things* in a totally illegitimate way when I speak of the evident non-simultaneity of *events*.

But let's see where the arbitrary definition of time takes us. Let us now define the two events as simultaneous. What strange consequences follow from this! At the moment when I knocked, for example at 1:25, a sound wave started and arrived at the point where you knocked before you started knocking. Before that? Yes, before. Because a clock would have registered my sound wave before your knocking. However, by virtue of our arbitrary definition of time, at the moment you knock, your clock only reads 1:25. So the sound wave started from me at 1:25 and arrived at you at 1:23. So it went back in time! Paradoxical!

But what is the harm in that? The sound wave came to you at the very moment you were lighting a cigarette. So you *always* lit it at 1:23 and these two latter facts remain linked together. Then the sound wave, reflected on your wall, came back to me. When it arrived I still had my hand on the wall; it was 1:25 and 1/10 of a second. It therefore took 2 min and 1/10 of a second to travel back. A very slow sound! The time, measured by me, for the outward journey and return together is 1/10 of a second. The return time is 2 min and 1/10 of a second. The calculation is correct, however, because the time for the outward journey was negative: $-$ 2 min. The sum therefore is equal to 1/10 of a second.

The numbers given by the measurements of velocities thus become somewhat strange. But in spite of the disturbance this causes to our habits of mind, it does not, and we emphasize this, bring any change to the content of our *perception*. You have perceived that the sound wave arrived at the exact moment when you were lighting the cigarette, and this exists even in this strange temporal order. *In the same way* the fact exists, which was perceived *by me*, that on its return the sound wave arrived while my hand was still on the wall. Therefore, the psychic limitation that leads me to recognize this coincidence is not in the least affected. We observe how our definition of time only concerns the *interpretation* of our perceptions and not the perceptions themselves. The immediate data are not questioned; changes are only made to the *order* in which these data ensure the coherence of the psychic world.

This is precisely why an aprioristic faculty cannot tell us anything about the legitimacy of our definition of time. What we actually experience is not disturbed by our changes made to the order of phenomena. The arbitrariness in the choice of simultaneity is so great that it *never* leads to a false claim about what has in fact been observed. This is why our observation of time is never false or contradictory to the psychic limitation of perception.

IV

However, it is extremely improbable that the speed of sound in a given direction becomes negative. Is that not impossible? Can a signal go back in time? We would gladly state that this is ruled out a priori.

Excluded? Yet that is what occurs! The sound wave goes back in time! We certainly have the right to say that what is real is also possible.

But if this sound wave propagates negatively in time, this is due solely to the inconvenient definition we have given of simultaneity. If we had defined it in the ordinary way, negative velocity would not have intervened. It is a wave that goes back in time only *in appearance*; in *reality* it moves forward in time.

Appearance and reality? Let us leave these concepts aside for a moment. However, let us agree that the cause of negative velocity is our inconvenient definition of time: in *this* case it could be avoided and we will prefer, rightly, the more natural definition of time. But is negative time *always* avoidable?

This is an entirely different question. Is it *always possible* to *consistently* define simultaneity in such a way that all signals sent, or all bodies in motion, are assigned positive time differences marked by the clocks at which they pass? This question should not be answered hastily. For if we answer positively, we affirm something very important about nature: something concerning the *order of causal sequences* in the world.

We have already addressed this question in a more extensive paper published recently.[3] We can resolve this question only with the help of an *axiomatics of the theory of time*: we need to discover the presuppositions of the affirmative answer. Whether such presuppositions are realized in nature is yet another question. It may be noted that they are realized in the theory of special relativity; Einstein's definition of time, which posits the speed of light as constant, is such at the same time as to extend the requirement of causality to the positive durations of all causal sequences. *Also* in the general theory such assumptions are normally realized, but it is still an open mathematical question whether they are so for all possible gravitational fields.

However, one thing is clear: it is not up to us to think about answering the question in the affirmative or not. Despite our reluctance to imagine that a state is possible in which the temporal order does not satisfy the requirement of causality, the answer to this question depends on the *facts* and not on *us*. The axiomatic formulation has made it possible to detect these facts with certainty: they concern the nature of the order of causal series in the world and can be determined independently of any definition of simultaneity.

Thus, the analysis of the problem of simultaneity leads to a surprising disjunction between *arbitrariness* and *necessity*. The definition of simultaneity is arbitrary, whereas the theory of absolute time regarded it as necessary. On the contrary, the implementation of a temporal order responding to causality does not depend on our choice; whether such an order is possible depends on empirical facts. At this point,

---

[3] Hans Reichenbach, *Axiomatik der relativistischen Raum-Zeit-Lehre*, Vieweg, Braunschweig, 1924.

absolute time theory believes the opposite, namely, that it is always possible to satisfy the causal requirement by comparison with the temporal order of cause and effect. It does not, however, note the axiomatic presuppositions involved in this question, the realization of which can only be established through experiment.

V

One last question remains to be resolved. It is a fact that for the implementation of the causal requirement there is no logical limit in the temporal order, to the point that a world is even conceivable in which it is not realizable. But *if* it is realizable, is a determinate definition of simultaneity not then prescribed with this requirement? Is there a way to eliminate relativity from simultaneity? Let us assume, if it is possible, that simultaneity is defined in such a way that all signals have positive duration. This is a legitimate requirement, the satisfaction of which gives the concept of time a very special meaning. Would not simultaneity be defined in this way univocally and without any arbitrariness?

This question, too, can only be answered by axiomatics. It says: in this way the relativity of simultaneity could disappear. But this in turn depends on the validity of certain axioms indicating the properties of causal sequences; and it is experiment that must decide whether they are valid. Let us thus formulate the axiom to be considered. If I send a signal from A to B and immediately afterwards a signal from B to A, then in A a certain time elapses between the departure of the first signal and the arrival of the second. Is it possible to make this time arbitrarily small through an appropriate choice of signal? Note that this question must be answered independently of any definition of simultaneity at distant points and independently of any time metric; it is a *topological axiom of time*. Depending on whether the answer is yes or no, simultaneity will or will not be free of ambiguity.

Experiment has already answered this question: given B, the time in A cannot be made small at will. The fastest signals are light signals, and they too take a certain amount of time to travel there and back. Faster signals do not exist, since experiments on electrons have shown that the kinetic energy of moving masses grows faster than quadratically with increasing speed and would even become infinite for the speed of light. This velocity constitutes an objective limit to any transmission of motion. Of course it is not asked that it be accepted a priori, but neither can its opposite be stated a priori. Experiment decides in favor of *one* of the possibilities. In this way, however, the possibility of univocally defining simultaneity is lost. This can be defined within a certain interval, while the causal requirement that all signals receive positive duration remains satisfied. The definitions of time that we gave in Sect. III by way of example are excluded: for domains as small as a room, the arbitrary interval is reduced to an imperceptibly small time. Thus, for such small domains, there is practically no difference between absolute and relative time; such a difference becomes observable only for astronomical dimensions or for systems moving at extremely high speeds, such as electrons.

VI

We have shown how the relativity of simultaneity is necessary from the point of view of the theory of knowledge, how it is not excluded by any psychic limitation, and how it is not suppressed by the requirement of causality either. Let us now examine the following question: what is the relationship between Einstein's time and everyday time? Can the order of phenomena, as considered by the theory of relativity, still be called time?

The concepts of everyday life are not formulated with sufficient precision to be directly comparable to scientific concepts. The term "time" has several meanings: First of all, we understand it as psychological experience. We perceive the passage of time, sometimes slower, sometimes faster; our consciousness represents it to us as a linear sequence. But in addition, we also always order the facts of the external world according to such a sequence; and the use of the clock indicates the replacement of the subjective temporal sequence by an objective mechanism, the functioning of which is seen as the order and measure of time.

In doing so, we always follow the principle of ordering time in such a way that the cause never follows the effect. This practically leads to an absolutely peculiar simultaneity because, as we said in Sect. V, the indeterminate interval is too small to be evidenced in practice. Thus the objective temporal order of everyday life has been established sufficiently unambiguously.

At the same time, however, the idea has taken root that this temporal order is the only one possible. Every primitive thought claims to be the only admissible one, a stage which only a scientific education permits one to overcome. Therefore only now do we notice by what peculiarity of causal sequences such a simple temporal order is realizable. This peculiarity is the high speed of light: because of this velocity, the time measured in A of the light signal ABA becomes extremely small and so does the admissible interval of simultaneity. If the limiting speed were much lower, for example, of the order of magnitude of the speed of sound, the arbitrariness in the definition of simultaneity, which the requirement of causality leaves unresolved, would have long since been detected. Even in everyday life, the regulation of clocks plays an important role; just think of the fact that every morning all clocks in railroad stations are regulated by signals. In our hypothetical case, the theory of relativity would have been invented long ago by a railroad engineer and there would have been no need for an Einstein. Today, no one would say that it is impossible to depict the relativity of time: it would have occurred all too often in practical life.

So let us also leave aside the distinction between appearance and reality. It does not belong to our argument. To pretend that only one of the definitions of time is valid and the other is only apparently valid is no more meaningful than it would be to assert that the true measure of lengths is given by the units of the metric system and that the use of other units leads to an apparent measurement of lengths. The scheme into which we make phenomena fit and into which we order them in order to describe them numerically is neither a reality nor an appearance. It is a conceptual system and, in observing whether it is applicable and unambiguous, a property of reality is expressed. The task of philosophy is to discover the presuppositions of such applicability, whereas it should refrain from defending common ideas. Eventually, in fact, these ideas will have to be oriented in the direction that a superior mind had indicated in advance. Just as today no one thinks that it is impossible to depict the spherical shape of the Earth, despite the fact that it cannot be seen, in the same way in the future no one will say that it is impossible to depict the relativity of simultaneity.

Correction to: Albert Einstein—Italian Memories

Correction to:
S. Linguerri and R. Simili, *Albert Einstein-Italian Memories*,
History of Physics, https://doi.org/10.1007/978-3-031-52950-4

In the original version of the book, the following correction has been incorporated in the Frontmatter: Affiliation of the second author has been changed from "Raffaella Simili (deceased), Bologna, Italy" to "Raffaella Simili, Department of Philosophy, University of Bologna, Bologna, Italy". The book has been updated with the changes.

The updated version of this book can be found at
https://doi.org/10.1007/978-3-031-52950-4

Appendix

Credits and Photos

1. Albert Einstein and Federigo Enriques in the loggias of the Archiginnasio, Bologna, October 1921. (By courtesy of the Historical Archive of the Zanichelli publishing house.)

S. Linguerri and R. Simili, *Albert Einstein—Italian Memories*, History of Physics, https://doi.org/10.1007/978-3-031-52950-4

2. Albert Einstein Portrait 1896. (By courtesy of the ETH-Bibliothek Zürich, Bildarchiv.)

3. View of the "Einstein, Garrone e C." building (ca. 1900) with the fall of water
   on the Naviglio Pavese that the company intended to use to generate electricity.
   This last project encountered difficulties and remained on paper. (By courtesy of
   the Musei Civici di Pavia.)

4. Federigo Enriques (third from the left) in 1911 at the University of St Andrews
    while receiving his honorary degree. (By courtesy of the Historical Archive of
    the Zanichelli publishing house.)

5. Gregorio Ricci-Curbastro (the first seated on the left) with his family. (By courtesy of Enrico Sangiorgi.)

6. Tullio Levi–Civita (the first on the right) and Vito Volterra. (By courtesy of Virginia Volterra.)

7. Tullio Levi–Civita Portrait. (By courtesy of the Accademia Nazionale dei Lincei and Biblioteca Corsiniana.)

8. Albert Einstein, Federigo Enriques and a group of professors from the University of Bologna in the Archiginnasio loggia, Bologna, October 1921. (By courtesy of the Historical Archive of the Zanichelli publishing house.)

9. Albert Einstein in the loggias of the Archiginnasio, Bologna, October 1921. (By
   courtesy of AIP Emilio Segrè Visual Archives.)

Author Index